How to Service and Repair Small Gas Engines

E. F. LINDSLEY

Drawings by Forrest J. Battles

Photographs by the author

Popular Science Books

SEDGEWOOD® PRESS, New York

Copyright © 1987 by E. F. Lindsley

Published by
  Popular Science Books
  Sedgewood® Press
  750 Third Avenue
  New York, New York 10017

Distributed by Meredith Corporation, Des Moines, Iowa

Library of Congress Cataloging-in-Publication Data

Lindsley, E. F.
  How to service and repair small gas engines.

  Includes index.
  *1. Internal combustion engines, Spark ignition—
Maintenance and repair. I. Title.*
TJ790.L56   1987   621.43'4'0288   87-6966
ISBN 0-696-11002-4

10 9 8 7 6 5 4 3 2

Manufactured in the United States of America

How to Service and Repair Small Gas Engines

Contents

Introduction   VII

Part I—Small-Engine Basics

1. The 4-Stroke Engine   3
2. The 2-Stroke Engine   9
3. The Tools You Need   15
4. Seasonal and Routine Service   24

Part II—Troubleshooting and Repair

5. Troubleshooting Starting Problems   41
6. Manual Starters   58
7. Servicing the Fuel System   69
8. Governors   97
9. Flywheels and Flywheel Brakes   110
10. Servicing Ignition Systems   120
11. Troubleshooting and Servicing Electrical Systems   146

Part III—Overhaul or Replacement

12. Should You Rebuild Your Engine?   175
13. Cylinder Heads   187
14. Valves, Seats, and Guides   199
15. The Crankcase   221
16. Crankshafts, Camshafts, and Oil Pumps   244
17. Pistons, Rings, and Connecting Rods   257
18. The Cylinder Bore   279
19. Choosing and Installing a New Engine   290

Index   307

It doesn't take long for a suburban household to accumulate a wide variety of lawn and garden tools powered by small gas engines. Sizes can range from mini two-stroke engines on hedge cutters to engines with two-figure horsepower on tractors, snowblowers, and other popular machinery.

Introduction

If you have a small gas engine that takes more than three pulls to start, this book was written for you. Maybe you've been dismayed by the high cost and slow service at engine-repair shops. Or, perhaps you want to keep a new lawnmower in top running shape, and are wondering about the whys and wherefores behind the manufacturer's instructions. If so, then this book is for you, too. It will show you how to tackle most repairs, even overhauls, yourself. It will also tell you when to back away from those few jobs that require highly specialized equipment.

Most of us, if we're suburbanites, have at least one and perhaps a dozen small gas engines powering everything from a rotary mower to a garden tractor, including chainsaws, tillers, weed trimmers, yard blowers, and maybe an emergency pump or generator. Within the past few years, the state of the art in small-engine design has advanced greatly. They've been down-sized and lightened remarkably for hand-held equipment. And, they've become much more sophisticated with solid-state ignition and built-in battery-charging. The latter is especially important for today's breed of electric-start mowers. Also, government-mandated safety regulations now require walk-behind mowers to have a blade brake or clutch—features unheard of a few years ago. This book explains how these important devices work.

I purposely divided the book into three basic sections. The first four chapters cover small-engine fundamentals, from the differences between two- and four-stroke engines to what you need to know to maintain and store your engine safely. Section Two—Troubleshooting And Repair—begins with a chapter on starting problems written especially for readers who are not small-engine experts, but would like to start saving on expensive repair bills.

Succeeding chapters deal in depth with each section and component of typical small engines. You're shown step-by-step how to handle your own repairs and what to anticipate along the way—right on down to complete disassembly and overhaul, covered in Chapters 12 through 18. You'll also find out when repairing may not be worthwhile, as well as what to look for when buying a new engine, covered in Chapter 19.

No single book could possibly cover all the nuts-and-bolts service details of all the engines on the market. In most cases there are readily available, inexpensive manufacturer's manuals that supply specific fit, clearance, and torque values, and describe any changes in parts. Though such information is important, these manuals are sometimes confusing and are written primarily for experienced mechanics. Besides, it is equally important that you know why the parts of an engine are built the way they are, what they do, and what their vulnerabilities are.

Fortunately, most small engines are not complex. All have the same basic components—cylinders, pistons, crankshafts, carburetors, and ignition systems. The principles of troubleshooting and repair are the same, and your knowledge will work for all of these engines. You may also want to rig an engine on a machine or piece of equipment of your own. Again, some basic engine knowledge will make it easier as shown in Chapter 19.

Finally, because professional repair service is so expensive, a few minutes of your own troubleshooting and a simple repair often will more than pay for the cost of the tools you need. Some jobs absolutely require professional shop equipment, or are simply beyond the skills of an amateur. When that's the case, I say so. But nine times out of ten you'll have your engine back running in less time than you'd take to haul it to a dealer's shop and back.

—*EFL*

SMALL-ENGINE BASICS

The Four-Stroke Engine

The first step to keeping your engine running smoothly and finding out what's wrong when it's ailing is to understand how gasoline is used in the combustion process. Most of the larger engines run on a combustion system called a *four-stroke cycle*. When you understand the basic four-stroke cycle, all else falls into place. Note that I call it a "cycle," because that is what your engine does—it repeats a cycle of the same four events over and over from the same starting point to the same finishing point. The four strokes in this cycle are intake, compression, power, and exhaust.

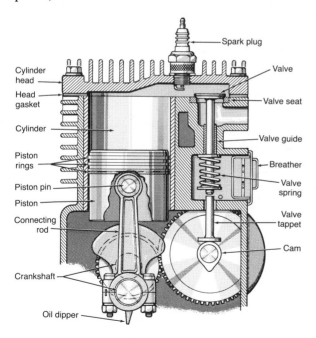

Working parts of all small four-stroke engines are basically the same. The crankshaft converts the reciprocating motion of the piston to rotating motion, which can be coupled to the load.

HOW THE FOUR-STROKE CYCLE WORKS

Four-stroke engines often vary in the mechanical arrangement of their parts, usually to make them better suited to the jobs they do. For example, it's obviously handier for the manufacturer of a rotary lawn mower to install an engine with a vertical crankshaft rather than build a right-angle drive for a horizontal crankshaft. In other cases, the engine manufacturer may elect to locate the intake and exhaust valves in the cylinder head for improved combustion efficiency. Nevertheless, the basic four-stroke cycle remains unchanged.

The Intake Stroke

Before your engine can run, it must first take in a combustible mixture of fuel and air. If you've never taken an engine apart, visualize the piston as a closely fitted, almost air-tight plunger that slides up and down in a finely finished cylinder. Picture the piston up near the top of the cylinder, a position commonly called Top Dead Center, or TDC. Now, imagine yourself grasping the connecting rod and pulling downward. You wouldn't move it very far before you felt the effects of atmospheric pressure, about 15 pounds per square inch (psi), trying to push the piston back up.

Because of this air pressure, the outside air wants to rush in to fill the void above the piston when the intake valve opens. Normally, the intake valve opens very near top center, and as the piston moves down on the intake stroke the air, carrying fuel vapor from the carburetor with it, continues to flow into the combustion chamber. Thus, the intake stroke is the first event of the four-stroke cycle.

A horizontal-shaft four-stroke engine. This basic type has powered reel mowers, tillers, small tractors, and many other machines through the years.

The rotary mower gave rise to the vertical-shaft engine. Auxiliary horizontal output shaft at base can drive mower wheels at slow speed while the crankshaft spins the blade at high speed.

This overhead-valve four-stroke engine gets more power because of better fuel mixture and exhaust flow. Valves, rocker arms, and push rods are housed in box atop engine. Note spark plug boot.

The Compression Stroke

When the piston is near bottom center, the intake valve closes. The combustible mixture is now trapped in the cylinder bore between the top, or crown, of the piston and the cylinder head. As the piston moves up, the fuel/air mix is squeezed or compressed into a much smaller volume. On many small engines, when the piston reaches TDC the volume will be only about one-sixth what it was at the bottom of the stroke. This is the second event in the four-stroke cycle.

The Power Stroke

Near the top of the piston travel, a spark from the spark plug is timed to ignite the fuel/air mix. In practice, this spark is normally timed so it occurs slightly before the piston is at the top. This is called *spark advance,* and is designed in because it takes a fraction of a second for combustion to get underway

and for the expansion of the burning gases to start pushing the piston downward. This is the third event in the four-stroke cycle. In your small engine, this expansion force may amount to over one ton of pressure applied to the piston many times each minute it runs.

The Exhaust Stroke

Since the fuel/air mix has now been burned, the exhaust valve opens as the piston nears bottom dead center on the power stroke. Then the piston moves upward again and forces out the waste gases. This is the fourth event in the four-stroke cycle.

Flywheel action. There are some important things to note about this sequence of events. First, only one of the working strokes, the power stroke, produces power. The other strokes are either getting ready for the power stroke or cleaning up after it. It takes two full revolutions of the crankshaft to get one power stroke. This means that a single-cylinder engine tends to run less smoothly than a multi-cylinder engine in which the power strokes more or less overlap to maintain a constant rotational force on the crankshaft. A relatively heavy flywheel is needed to keep the engine turning during the three non-power

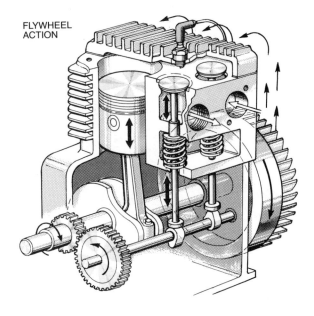

FLYWHEEL ACTION

The rotating mass of the flywheel stores some of the energy from each power stroke.

The Four-Stroke Cycle

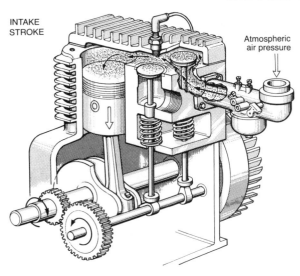

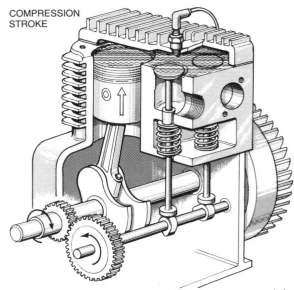

1. When the camshaft lifts the intake valve open, atmospheric pressure forces air through the carburetor where it picks up gasoline vapor. Valve remains open until piston stops its downward travel and starts back up on compression stroke.

2. Upward movement of the piston compresses air/fuel mixture into roughly one-sixth the volume that existed at the bottom of the piston stroke. Little power would be produced if the mixture were not compressed. Both valves are closed.

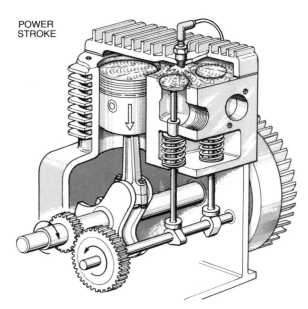

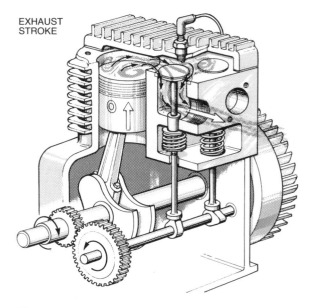

3. Just before the piston reaches the top, a spark from the spark plug ignites the air/fuel mixture. The heat causes rapid expansion of the air and gases against the piston, driving it down again on the power stroke. Both valves are closed.

4. The exhaust valve opens near the bottom of the power stroke. At the top of the exhaust stroke, the spent gases will have been pushed out and the cycle is ready to begin again. At 3,600 rpm, these events are repeated 1,800 times each minute.

Oil level

Operating a four-stroke engine at an extreme tilt-angle may cause momentary loss of lubrication as shown.

strokes. The inertia of the spinning mass overcomes engine friction, and the workload of pumping in air, compressing it, and expelling the burned gases. In addition, the flywheel has to force the camshaft to turn against the compression of the intake- and exhaust-valve springs. If an oil pump is used, it represents another load that must be carried by flywheel inertia. So do the cooling fan or blower blades on the flywheel itself.

ESSENTIALS FOR COMBUSTION

A four-stroke engine must have certain components for combustion to take place. Although these parts may appear somewhat different on different engines, they are basic to all engines of this type and include:

- A tightly sealed combustion chamber in which the fuel and air can be compressed and burned.
- A carburetor to meter the fuel, vaporize it, and mix it with air.
- An intake valve to let the fuel/air mix into the combustion chamber and then seal it there.
- A spark to ignite the fuel/air mix.
- An exhaust valve to allow the burned gases to escape so a fresh mixture can come in.
- Timing mechanisms to open and close the valves and time the spark.

What is often more difficult to envision is just how incredibly fast the four timed events occur. A typical small four-stroke engine may run at 3,600 revolutions per minute (rpm). That is, the crankshaft will turn 3,600 revolutions in one minute, which means there must be 1,800 explosions—30 per second—in

the combustion chamber. Yet with a minimum of attention and maintenance from you, your little engine will provide many hours and many seasons of hard work.

Four-Stroke Characteristics

The four-stroke engine attached to your mower, tractor, or other machine has certain advantages, as well as disadvantages, that make it different from the two-stroke engine on your chainsaw or weed trimmer. It's important to understand some of these before you buy a piece of equipment in cases where you have an option of four- or two-stroke power.

Lubrication-system reservoir. Since there is no oil introduced with the fuel in a four-stroke engine, a separate oil reservoir is needed at the base of the engine. Also, some form of oil pump or other device must be provided to deliver oil to the working parts. If you hate the bother of measuring and mixing oil and fuel, the four-stroke engine is more convenient.

On the other hand, if the equipment the engine is powering must frequently operate at an unusual angle—for example, a rotary mower on a steep slope—the oil in the engine base may be displaced so that the oiling device can't pick it up properly. Obviously such a system won't work at all if you want to invert the engine or tip it sideways, as you would on a chainsaw. Another disadvantage arises if your engine powers a generator or pump that must run continuously. You may simply run out of oil and ruin the engine.

Valve train. The four-stroke engine is also fairly complex, requiring both an intake and exhaust valve

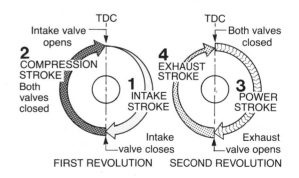

In practice, intake opening and exhaust closing often overlap a few degrees.

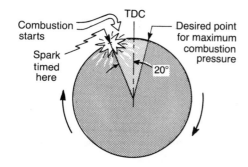

Most ignition timing is set so the spark occurs just before the piston reaches top center. The timing advance allows combustion to develop and the resultant pressure to build so maximum pressure is reached just after top center.

that seat and seal tightly, camshaft, cam gear, and lifters to open the valves, and valve springs to close them. These make a four-stroke slightly more expensive to produce and repair than a two-stroke simply because it has more parts.

A timed valve opening and closing also means that the cycle of events is accurately controlled even when other parts of the engine are worn. Everyone has probably seen faithful old four-stroke engines that use oil and smoke, yet keep putting along. Two-strokes tend to be less tolerant of severe wear.

High torque. Still another advantage of the fixed, accurate timing of events in a four-stroke engine is low-speed torque or lugging ability. Most people look at engine horsepower ratings, but horsepower and torque are two different things. Horsepower refers to the *rate* at which an engine can do work when running at an optimum speed. For example, a 4-horsepower engine can perform a certain amount of work at 3,600 rpm. But when the load lugs the engine, as when tall grass slows your mower or deep snow your blower, it's really torque—the ability of the crankshaft to put out twisting effort without stalling—that keeps you going. The valve action of the four-stroke engine usually provides better torque and load handling than can be mustered by a two-stroke of equal horsepower.

Horsepower And Torque: Four-Stroke Vs. Two-Stroke

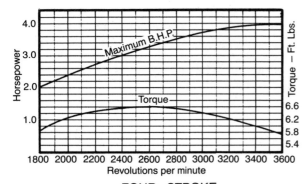

FOUR - STROKE

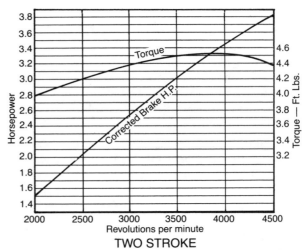

TWO STROKE

Horsepower on a typical four-stroke, 4-hp engine peaks at 3,600 rpm. As load drags speed down, torque or twisting force builds up to keep the engine going under heavy variable loads.

On a typical 4-hp, two-stroke engine, horsepower peaks way up at 4,500 rpm. When slowed by load, torque starts to fall off and the engine will slow down even more or stall.

The Two-Stroke Engine

The two-stroke engine fits many power-equipment needs because it is simpler than a four-stroke, has fewer parts, can be made lighter, and most important, can run in any position. Chainsaws, small grass and hedge trimmers, and the like would be impractical with a four-stroke engine in most cases. Nevertheless, the same essentials for combustion exist:

- A tightly sealed combustion chamber.
- A carburetor to measure fuel and mix it with air.
- Some means to admit fuel and air.
- A spark to ignite the mixture.
- An exhaust path for the burned gases.

HOW THE TWO-STROKE CYCLE WORKS

The term two-stroke means that the two-stroke engine accomplishes all the functions of the four-stroke cycle with only two strokes of the piston, one up and one down. Such an engine fires every revolution of the crankshaft. Clearly, the events occurring in the cycle must be arranged differently than for the four-stroke, even though the essential results of expanding the gases against the piston remain the same.

Two-Stroke Intake

Not only is oil mixed with the fuel in a two-stroke engine, the fuel/air charge does not go into the combustion chamber directly but first enters the crankcase. In the two-stroke engine, picture the beginnings of intake with the piston at *bottom* center. The carburetor is connected to the crankcase. As the piston moves upward it leaves a void, this time *below* the piston and in the crankcase. In the simplest two-stroke, a number of flexible, springy reeds—actually thin strips of metal—are mounted on a reed plate between the carburetor and crankcase. The reeds

This little engine is typical of the extreme light weight common to two-strokers.

can only open under suction and are forced away from their flat seating surface on the reed plate as the piston goes up, letting the fuel/air mix rush in under atmospheric pressure.

Transfer Action

At this point we have a combustible charge in the crankcase, not the combustion chamber. It must somehow be transferred to where it can burn. Now picture a transfer passage, or bypass, leading from the crankcase to a hole or port in the cylinder wall.

As the piston comes down, the fuel/air mixture is trapped in the crankcase because the reed valves only open one way and are now closed. Crankcase pressure builds and the fuel charge seeks an escape

9

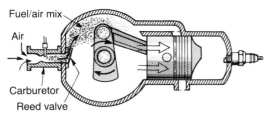

REED VALVE TWO-STROKE

Unlike a four-stroke, a two-stroke engine draws air and fuel into the crankcase as the piston rises toward the combustion chamber and leaves an area of reduced pressure behind it.

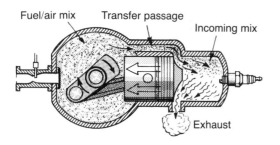

The fresh fuel/air mix must enter the combustion chamber with enough velocity to help move out the burned gases, yet must also remain in the combustion chamber to be compressed when the piston moves up. Some fresh fuel is always lost through the exhaust port but careful design holds it to a minimum.

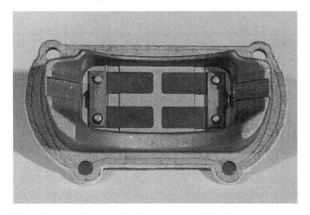

Four dark rectangles in the center of this reed plate are springy, flexible reeds that cover the openings through which the fuel/air mix enters the crankcase.

At or near piston top-center, the spark plug will ignite the fuel and the piston will be driven downward until the exhaust and transfer ports are cleared by the piston.

path, which suddenly becomes available when the descending piston uncovers the port in the cylinder wall at the top end of the transfer passage. The fuel/air mixture instantly pops through this port and into the combustion chamber.

Scavenging

At about the same time the fresh fuel/air mix is entering the combustion chamber, the piston has also uncovered an opposite port in the cylinder wall called the exhaust port. Much of the exhaust will escape under its own residual expansion force from the power stroke, but some of it will also be driven out or "scavenged" by the inrushing fuel/air mix.

Compression

As the piston reverses at bottom center and starts up, it will cover both the transfer port and the exhaust port to trap and compress the mixture between the piston crown and the cylinder head. Thus, at the same time the engine is compressing and preparing to ignite the charge it is also taking a fresh charge into the crankcase. And after ignition, as the

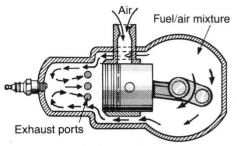

LOOP SCAVENGE WITH THIRD PORT

Many two-strokes use piston-porting to introduce fuel and then seal the crankcase. Here, the piston covers and uncovers a third port in the cylinder wall for a timed entrance of the fuel/air mix.

piston starts downward on the power event, it is also compressing the mixture in the crankcase in preparation for its transfer to the combustion chamber. This means that there is compression at both ends of a two-stroke engine, in the crankcase and above the piston.

Baffle and loop scavenging. Note that during the time the transfer port is open to allow a fresh fuel/air mix to enter, the exhaust port is also open. That means some of the fresh charge may cross the top of the piston and be wasted out the exhaust port. Many different techniques are used to prevent or at least minimize this waste. The simplest is a little wall or baffle on top of the piston to block the direct cross flow of the fuel/air mix. Alternately, the piston crown may be contoured to direct the mix upward into the space above the piston. This basic strategy is still fairly common, though some makers have improved performance by positioning, angling, and contouring multiple transfer ports that cause the incoming charge to loop upward towards the top of the combustion chamber in a carefully controlled pattern. This is called *loop scavenging.*

Reed valves and piston porting. Although we used the reed valve system in our explanation of the two-stroke intake event, the same opening and closing action can also be accomplished with ports that are covered and uncovered by the piston. Such engines are called *third port* or *piston-ported* engines. Still another form of porting makes use of a flat spot on or hole in the crankshaft, which opens and closes a pathway for intake air at certain positions of crankshaft rotation.

Transfer port

Here the piston has moved down and opened the exhaust port and both transfer ports. The latter are angled to direct the mixture upward in a swirling pattern.

Four-stroke mufflers such as the one held in hand are almost generic. Nearly any one of the right size will do. In contrast, the dark violin-shaped object below it is a carefully tuned two-stroke muffler with interior passages matched to engine characteristics. Thus, a replacement must be the exact one specified.

TWO-STROKE CHARACTERISTICS

Think of the reasons given for a heavy flywheel on a four-stroke engine and you'll see that with the two-stroke firing on every downstroke there is less need for flywheel inertia. The loads of the camshaft, valves, and oil pump are also gone. That means substantial weight and bulk represented by the flywheel can be omitted from a two-stroke engine. The flywheel on a chainsaw engine, for example, is barely large enough to house the ignition parts.

Noisier running. If your two-stroke engine is running at 5,000 rpm you have 5,000 firing impulses each minute, compared to 2,500 for a four-stroke at the same rpm. Consider the doubled frequency of the exhaust "pops," plus the opening of the intake port simultaneously with the exhaust port, and it's understandable why the exhaust sound of a two-stroke engine is immediately recognizable and distinct from that of a four-stroke. It is also much harder to muffle a two-stroke engine effectively since any back pressure against the existing exhaust gases leaves some unscavenged gas and impairs performance. Remember, in a four-stroke engine the piston pushes the exhaust gases out, but only residual gas energy and the incoming charge are available to get the waste gas out of a two-stroke.

While almost any exhaust pipe and muffler of adequate size will do for a four-stroke engine, be assured that the engineers were quite careful about the muffler design on your two-stroke, and you're well advised not to try and outguess them. In addition, many small engines—particularly chainsaws—have special spark-limiting mufflers, sometimes referred to as Forest-Service Approved. Be sure to replace such mufflers with the same approved type, preferably from the equipment or engine builder.

Fuel-and-oil lubrication. You can twist, turn, invert, and shake a two-stroke without worrying about oil sloshing away from vital internal parts, since the oil is mixed with the fuel. There are no valves, camshaft mechanism, and associated valve-train parts to heft. And since firing takes place on every other stroke, you do get more power from the same package and therefore the package can be made smaller and lighter.

High rpms. Without valves to open and close, it is also possible to run a two-stroke at a much higher rpm—often as high as 8,000–11,000 rpm—again, producing needed power from a smaller, lighter package. Also, the two-stroke responds very quickly to a twitch of the throttle, in part because a light

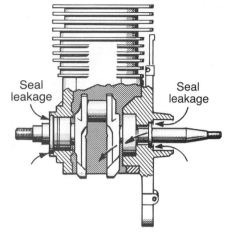

Seal leakage   Seal leakage

A two-stroke's crankcase seals are vital to easy starting and powerful performance. Don't overlook other possible leaks between the halves of a split case or through the reed valves.

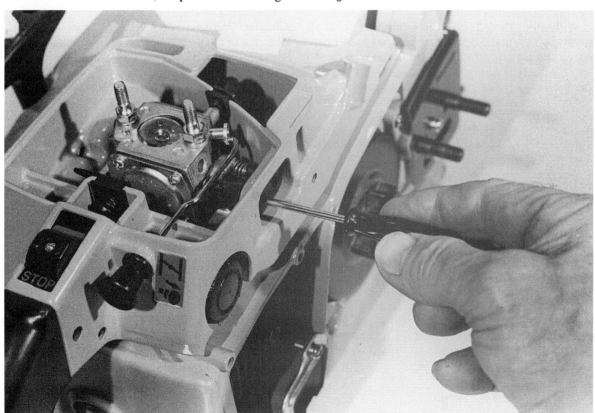

To adjust a two-stroke's carburetor, it helps to first remove the air cleaner so you can locate the screws.

Replace the air cleaner before starting and adjusting. Be sure to count turns and half-turns.

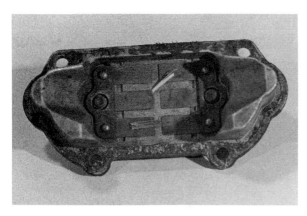

Good air cleaner maintenance is also especially important on two-strokes. Even tiny bits of grass or sawdust in the reed valves will hinder starting and cut the engine's power severely.

flywheel accelerates faster than a heavy one.

With all of those features it would seem that all small engines should be two-stroke. That hasn't happened and isn't likely to. For one thing, many owners object to the mess and bother of mixing oil with the fuel.

Poor low-end torque. More important, most small two-stroke engines depend on those high rpms to produce their rated power; when they're lugged down under heavy load they simply stall. Unlike a four-stroke with its timed valve action, the two-stroke needs extremely high fuel-charge velocities as the mix moves through the ports. Designers tune the porting size and locations for certain optimum speed ranges, and when the engine speed drops below this design range the tuning just isn't right.

Carburetor adjustment is critical. If you turn the main carburetor adjustment screw towards lean on a typical four-stroke engine, the engine will simply start to slow down or pop back. Try the same adjustment on a two-stroke, and as you get leaner the exhaust will suddenly assume a high, screaming tone as the engine speed increases. This may appear to be the best adjustment, but it can actually damage your

engine quickly by causing piston scoring and detonation due to overly high combustion temperatures.

Plug fouling. Unlike four-stroke engines, whose piston rings are designed to keep excess oil from the combustion chamber, two-strokes can often develop spark-plug fouling because the oil is carried up toward the plug along with the fuel. Moreover, the wrong fuel/oil proportions, improper mixing, or use of an oil not intended for two-stroke engines (such as automotive oil) can quickly foul a plug and prevent running.

Plug fouling can also result if an attempted start or a low-speed or brief run leaves a puddle of unburned fuel in the crankcase. The gasoline portion may evaporate and leave a residue of heavy oil, which can find its way through a transfer port and onto a plug. This can happen if you carry a small piece of equipment such as a chainsaw in a car trunk positioned so the residual oil can run into the combustion chamber.

Crankcase sealing is critical. I mentioned in Chapter One that even a badly worn four-stroke engine will often run and run even though the bearings are dribbling oil around the crankshaft. This is seldom true of a two-stroke. You'll recall that to get fuel and air into the crankcase of a two-stroke you must have a partial vacuum or negative pressure created by the upward movement of the piston. If the crankshaft bearings and seals are worn, this negative pressure leaks away; air enters around the seals and defeats the intake process. Whatever fuel that does get in tends to leak out around the seals before it can be compressed enough on the piston's downward stroke to blast through the transfer port into the combustion chamber.

Another potential source of crankcase leakage are defective or fouled reed valves. If you are careless with the air cleaner on a two-stroke, a minute fragment of sawdust or grass can wedge a reed valve off its seat and you'll have great difficulty starting the engine. Remember that because compression takes place at both the top and bottom end, the crankcase of a two-stroke engine must seal as well as the combustion chamber.

3

The Tools You Need

For a year or two, while your engine is new, a few simple tools are normally all you need for routine service. You probably have suitable regular and cross-slotted screwdrivers, pliers, and maybe enough wrenches for checking the screws, nuts, and fittings that might loosen up. For the small external sheet-metal screws and other fasteners, there is nothing handier than a set of ¼-inch-drive sockets ranging downward from ½ inch to ³⁄₁₆ inch. You'll find them much better than larger wrenches because of the limited access on small engines, particularly chainsaws and trimmers. Sockets also reduce the chance of overtightening and stripping the threads on small fasteners. Be aware, though, that there's a good chance the nuts and bolts on your engine may be metric and require metric wrenches.

A set of ¼-inch-drive sockets, extensions, and ratchet will more than pay for itself when servicing your engine. They can get in neatly where large wrenches are often awkward and clumsy.

TOOLS NEEDED FOR ROUTINE SERVICE

Small to medium-size conventional and cross-slot screwdrivers.

Set of ¼-inch-drive sockets, ½- to ³⁄₁₆-inch; check for metrics.

Set of small to medium, ³⁄₈- to ⁹⁄₁₆-inch open-end and box wrenches.

Spark-plug boot puller.

Spark-plug socket. Check plug hex size; it may be ¹³⁄₁₆ inch, ¹¹⁄₁₆ inch, ¾ inch, ⅝ inch, or 14mm.

Wire-type feeler gauge and spark plug electrode bender.

Bulk-storage gasoline container, as well as dispensing container to deliver fuel to engine tank.

Mixing vessels for two-stroke fuel and oil.

Small offset funnel for filling four-stroke crankcase.

Thumb-pump oil can.

Used-oil storage containers.

Troubleshooting and Minor Repair

Usually an engine owner can do far more than he realizes, but what may be an easy job to one person may appear quite formidable to another. Some examples of troubleshooting and minor repair include:

- Removing, cleaning, and replacing the carburetor or installing a float needle and seat. This is almost routine if gasoline is left in the carburetor during long storage.
- Removing the blower housing and cleaning out debris and grass from the cooling fins. This may require retorquing a few cylinder head bolts.
- Replacing the recoil starter clutch, spring, or rope. May require a large socket, flex handle, and torque wrench to reassemble.
- Adjusting or replacing breaker points. May re-

quire a flywheel puller and torque wrench for reassembly.
- Fuel pump removal and rebuild—usually a simple job.
- Replacing cylinder head gasket. Not difficult but requires a torque wrench.

TOOLS NEEDED FOR TROUBLE-SHOOTING AND MINOR REPAIR

Tools previously listed, plus—

Socket set, ⅜-inch-drive, ⅜- to ¾-inch sockets, flex handle, and ratchet handle; for engines over about 4 hp, include ½-inch-drive sockets for cylinder head bolts and flywheel.

Torque wrench, ½-inch-drive for most four-stroke engines, ⅜- or ¼-inch-drive for chainsaws and small two-stroke engines.

Flywheel puller, universal or knock-off type.

Flywheel holders to prevent rotation when removing or installing flywheel nut; 3 feet of sash cord or nylon rope will do.

Two- and three-jaw and back-plate pullers to remove V-belt pulleys and other devices from the engine's crankshaft.

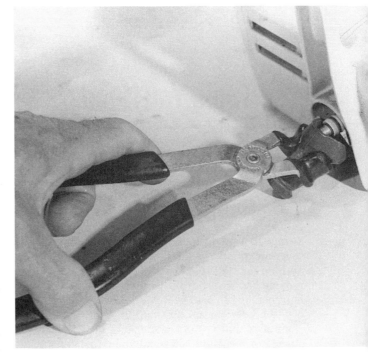

Tugging a plug wire loose, especially when the connector is down in a well, can cause trouble. A boot puller sneaks it out easily without damage to the connector.

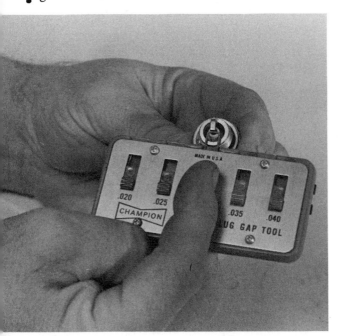

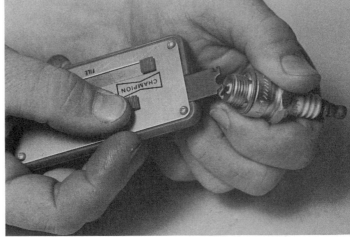

Be sure to use a wire gauge, not a flat one, to check spark-plug gap. The wire should just push through the gap without forcing. Adjustments are made by gently bending the outer electrode (above). Never pry against the center electrode.

A variety of fuel containers are available. The large flat container at left-rear is for remote connection and delivers fuel directly to the engine. Others are for bulk storage and dispensing fuel into engine tanks.

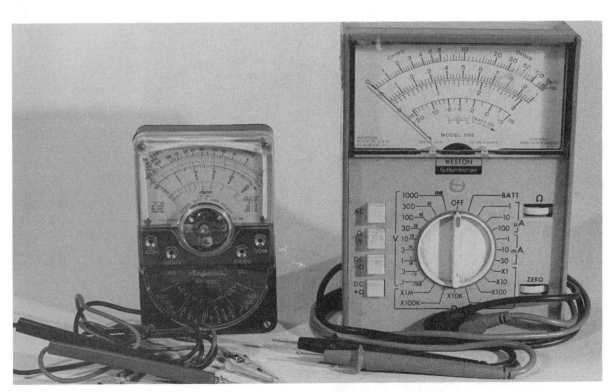

Volt/ohmmeters, commonly called VOMs, come in many shapes and forms. They're extremely handy for troubleshooting ignition and electrical problems. No home workshop should be without one.

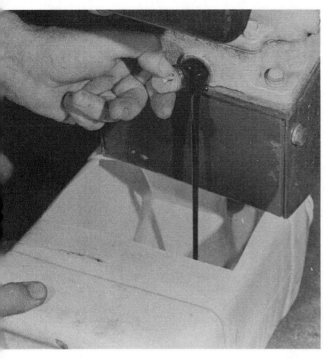

A 10-quart container cut open as shown above has a screw top you can open to pour dirty oil into a disposal jug. For badly placed drain plugs, add a length of pipe, below, to get the oil neatly into a drain pan.

Hex-key wrenches, for removing Allen screws from pulleys.

Helpful Additions include—

Volt/ohmmeter to check ignition and charging circuits; be sure to get at least the 20,000 ohms/volt type.

Compression gauge.

Snap-ring pliers; get the type that adapt to both internal and external rings with changeable tip sizes.

Flat feeler-gauge set.

For valve and seat repair you'll also need a valve-spring compressor, especially for engines above 4 hp. If you remove the valves and take the engine to a shop for seat and valve refacing, no other tools are needed. To reface seats yourself, add a carbide cutter and pilot kit to the list. New valves are usually cheaper in the long run, however.

Your decision to do any of the common jobs listed also depends on first checking a shop manual specific to your engine and making your own evaluation of the tools needed.

Heavier tools. There are certain tools you'll need if you want to disassemble your engine or rebuild a used one. For larger engines up to 18 or 20 hp in a tractor, you'll need ½-inch-drive wrenches plus larger box and end wrenches. These basic tools are a lifetime investment. You'll also need them if you do your own maintenance on a riding mower or tractor.

Torque wrenches. A torque wrench tells you how hard you're tightening a cylinder-head bolt, connecting-rod bolt, or flywheel retaining nut. Without a torque wrench, any serious repair work you attempt can cause major damage. For example, tightening the head bolts unevenly can distort the cylinder walls and cause piston and cylinder scoring. Many old-time mechanics claim they can pull the wrench to the proper tension by feel. *They can't!*

The torque wrench needed for larger engines 4-hp and up will be too large for chainsaw engines. Study the shop manuals for those engines in the size range you'll be working with and choose a wrench size and torque range accordingly. Larger wrenches may have torque scales in foot-pounds. This is too coarse a scale for small engines, and an inch-pound scale is easier to read.

Flywheel pullers. The flywheel is normally a very tight fit on the tapered crankshaft and won't come off without a puller. Some shop manuals show a threaded cap-like protector, called a *knock-off puller,* which threads onto the end of the crankshaft. By striking this tool a sharp hammer blow, the flywheel will usually be freed.

On the inexpensive tachometer at left, the little wire vibrates vigorously when you turn the dial to the engine speed. The electronic digital tach at right picks up a signal from the ignition system when held near or clipped onto the plug wire. Either unit will come in handy time and time again.

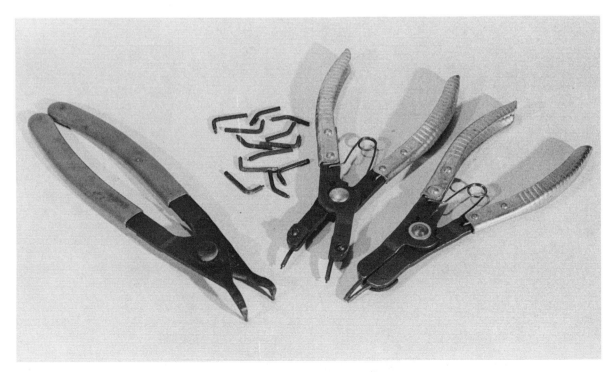

Removing and replacing snap rings can be difficult without the right pliers to spring the rings open or closed. The tool at the left fits only one size; those on the right have a selection of tips.

Some torque wrenches register on dials, some have pointers, while others have a built-in adjustable click that signifies proper torque. Scale on this pointer-type reads in inch-pounds; for foot-pounds, divide by 12.

Universal flywheel puller above has slotted openings to fit a wide range of flywheels with different puller-screw spacings. The three screws are inserted and tension is applied to the central jackscrew (left).

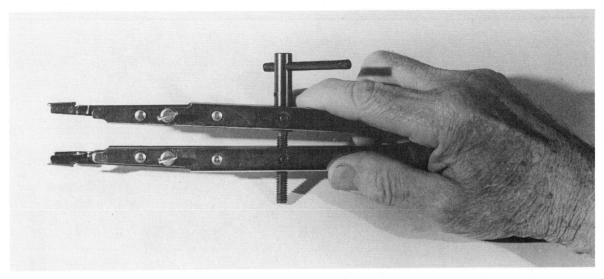

Valve-spring compressors certainly make removing or replacing valves easier. The tool above grips the spring at top and bottom and squeezes it, while the one below simply levers it up.

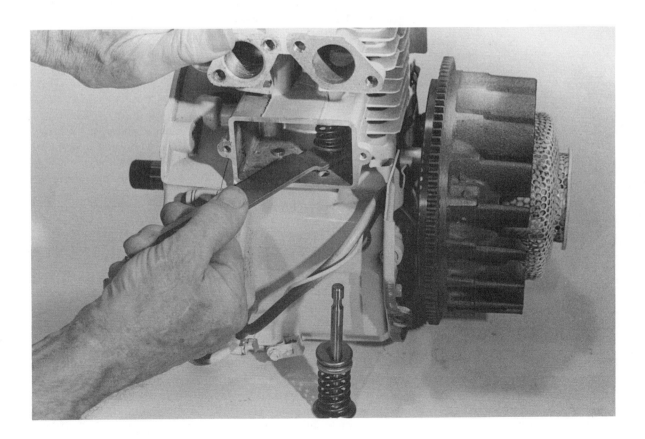

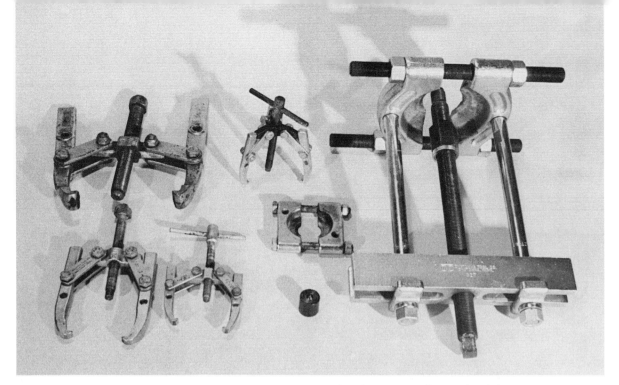

Jaw pullers at left should be used with the backing plate—center—to avoid breaking pulley flanges. The husky plate-type puller at the right is a necessity for pulling shrunk-on ball and roller bearings.

Knock-off flywheel remover is screwed on until it almost contacts the flywheel face. A sharp rap with a hammer will usually pop the wheel loose on the tapered shaft.

There are two types of *jackscrew flywheel pullers* commonly used. If the flywheel has three small holes around the hub, you can use the first type, a so-called universal puller. Such a puller has a heavy flange that seats against the wheel and a threaded center with a husky jackscrew. The flange has slots to adapt to different hole spacings in different wheels.

Many flywheels have only two puller holes and cannot be removed with the universal puller. These require a bar-like, two-hole puller available from engine parts dealers. It is also easy to make one from a bar of steel about 3 to 4 inches long, ½ inch thick, and 1 inch wide. These pullers simply bridge the end of the crankshaft, and the pulling is done by careful tightening of the screws into the flywheel.

V-belt pulley pullers. To work on almost any engine you have to remove V-belt pulleys, mower-blade hubs, and other parts that may be corroded in place or hung up on small burrs on the shaft. The die-cast or cast-iron flanges on such pulleys are easily broken by applying a two- or three-jaw puller directly. You'll need a small split puller-plate that fits over the shaft and behind the pulley to distribute the jackscrew force evenly on the pulley surface.

V-belt pulleys on older equipment can be extremely stubborn. Try soaking a little penetrating oil around the hub. Wait a minute or two to give the penetrant a chance to work; then try again with the puller as before.

TOOLS NEEDED FOR REBUILDING

If you intend to do more extensive repairs inside the engine, you'll need a more extensive list of tools. These include all of the tools previously listed, plus—

Ridge reamer, for removing the wear ridge at the top of the cylinder.

Measuring tools. These include 1-, 2-, 3-, and 4-inch outside micrometers, or a vernier caliper with a dial readout in thousandths of an inch; inside micrometers, 1 inch with extensions to 4 inches; dial indicator; machinist's scale.

Cylinder tools, with expandable hones of 1½ to 4 inches.

Reamers in valve-guide sizes and oversizes.

Seal-installing tools and seal-lip protectors.

Pullers, inside and out for ball, roller, and sleeve bearings.

Piston-ring expander.

Ring-groove cleaning tool.

Piston-ring compressor.

Wash tank, air compressor, arbor press, as well as any special manufacturer's tools needed.

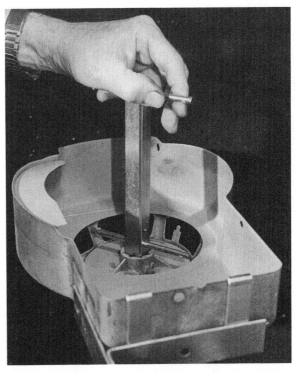

This specialty tool holds the blower housing firmly in a vise while you tighten the recoil starter spring.

23

4

Seasonal and Routine Service

In most areas your small-engine usage will reflect the season of the year, ranging from spring and summer yard and garden work to fall and winter chainsawing and snow removal. Yet, an engine standing idle can be harmed more in 30 days than after a full season of running. Ideally, once a week or so during the winter we would start up our mowers, tillers, and the like, and warm them up thoroughly. Then we would store them in a dry place after running the carburetors dry. Few of us have the time or inclination to do that, however, so we must find other ways to protect against storage damage such as:

• Crankshaft and bearing corrosion from acidic oil.
• Piston corrosion and ring sticking from old oil and fuel gum.
• Cylinder rust from lack of protective oil coating.
• Stiffened lip-type seals.
• Valve-stem gumming and valve-spring rust and corrosion.
• Carburetor gumming from stale gas.
• Carburetor and governor/throttle linkages gummed and sticking.
• Electrical parts and ignition damaged by moisture and oxidation.

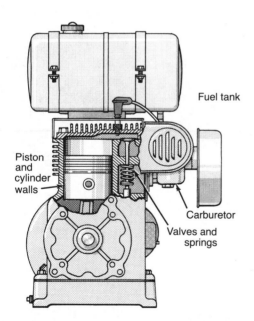

Fuel tank

Piston and cylinder walls

Carburetor

Valves and springs

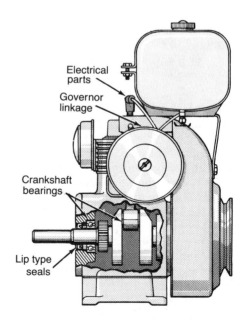

Electrical parts

Governor linkage

Crankshaft bearings

Lip type seals

Two views of this snowblower engine show where the ill effects of poor storage can concentrate. These same parts are similarly vulnerable on most other small gas engines.

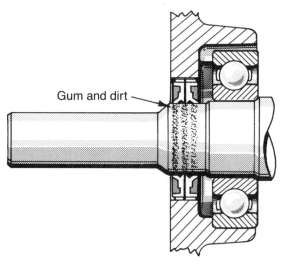

Gum and dirt

Long storage can cause crankshaft-seal lips to harden or accumulated dust and oil gum to solidify (above) so it abrades the seals or even wears a groove in the shaft where the lips contact.

STORING FOUR-STROKE ENGINES

The following procedures can protect your four-stroke engine from the types of damage listed. Probably the best way to do this is to bring together all of the equipment being stored so you can overlap operations. Some steps such as replacing the air filter and cleaning the cooling fins, though not directly related to storage, will prevent trouble next season. Slightly different instructions apply to two-stroke engines, covered later.

Clean the cooling fins. One of the most critical cooling areas on a four-stroke engine is around the exhaust valve, where exhaust-gas temperature can reach 1,400°F. Because the engines discussed in this book—both four- and two-stroke—are cooled by air circulated through the engine, the cooling fins serve a very important purpose. Ideally, you would remove the blower housing for access to the fins and the backside of the flywheel for cleaning. Do this if you have only to loosen a few screws or minor retaining bolts. On some engines, however, the housing is retained by two or more cylinder head bolts.

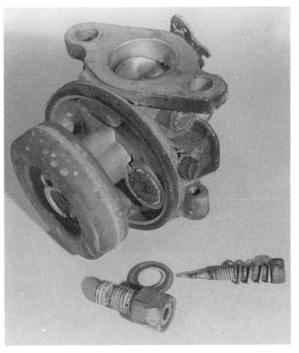

Gum, corrosion, and rust ruined the internal parts of this carburetor. The water came from the gas tank, the gum from old gasoline left in the carburetor.

A thorough poking with a piece of wire usually removes baked-in chaff and cuttings between the cooling fins.

Instructions for sponge-type filter elements frequently advise washing in detergent, rinsing, and squeezing dry. Some elements also require oiling.

In such cases it's wise to think twice. When the head bolts were originally torqued down they squeezed and compressed the head gasket. If you loosen them now and then reinstall them—even to the proper torque—the gasket may fail because it has hardened and lost its original elastic properties after many hours of running.

Before removing those critical bolts, try poking a wire down between the fins under the housing. Soft aluminum welding rod works well but a piece of coat-hanger wire will do. If you feel a lump of packed debris in the fins but can't get it out, you may still avoid removing the housing if you can get the machine to a source of compressed air capable of 100 psi pressure. Try blowing and poking to clean out the debris.

Clean or change the air filter. Many small-engine air filters are merely sponge-like foam pads. Often the instructions recommend washing them by squeezing in detergent or kerosene. Some foam or metallic filters also require oiling. Others are pleated paper or similar material and must be replaced. Even if the directions recommend cleaning, be skeptical of any filter that has seen several years of use. These tend to distort, lose their shape, fail to seal around the edges, or simply become clogged and ineffective. Replacing them costs a lot less than replacing a worn cylinder and piston.

Replace the spark plug. If the spark plug in your engine has a full season of service, replace it now. Check first in your manual for recommended plug reach, heat range, and whether it is radio-interference suppressed. If the plug seems to bind when removed or replaced, use a spark plug thread-cleaning tool to remove deposits from the threads in the head. You'll find detailed spark-plug service techniques in Chapter Five.

Check the manual starter. Pull the rope out to its full length and examine it for signs of fuzziness or abrasion. If the rope is worn, you can either remove the starter unit and take it to a dealer for service or follow the replacement procedure outlined in Chapter Six.

Also check the starter-clutch action. When you pull the rope you should feel the clutch or other

When inspecting or replacing disposable, pleated-paper filters, be sure the sealing surfaces at center and ends are clean and undamaged.

Dirt buildup on this two-stroke's air filter eventually became so heavy the engine could hardly run. Don't let yours get into this shape.

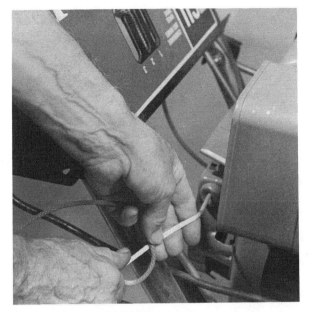

Pull out the rope on your recoil starter and go over it inch by inch. Look particularly at the part you seldom see—the clean area here.

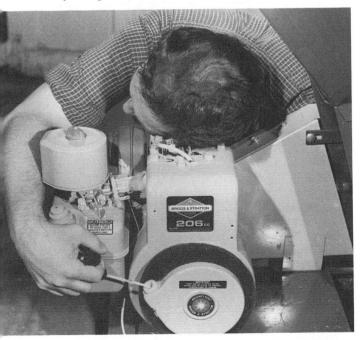

When your engine is new, try placing your ear against it and turning the flywheel. As time goes by, you'll pick up any unusual rattles or grinding that suggest a repair.

starter drive start to engage and turn the flywheel without slippage or erratic action. Such conditions tell you it's time for a new clutch before next season.

Manually rotate the engine. While you're checking the starter, place your ear down next to the cylinder head and rotate the engine by hand. Listen for sounds of scraping or grinding, which may indicate bearing trouble. Listen at the exhaust and intake for hissing sounds from leaking valves. Also try rocking the crankshaft back and forth and listening for clicking or clunking sounds that could come from a loose connecting-rod bearing or piston pin.

Inspect and oil the controls. Bowden control cables—which look like small-diameter springs and have the actual control wire running through them—often rust and become sticky or immovable. Before storage, brush or spray the outsides with kerosene mixed with light oil. Work the choke and throttle controls several times to lubricate the inner wire. If you detect limited movement or none at all, the cable may be kinked and need replacing.

At the same time, check the little anchor clips that secure the cable at the engine. A loose clip can let the cable slip and prevent full movement, so that the

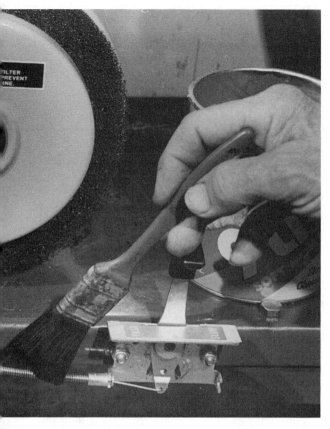

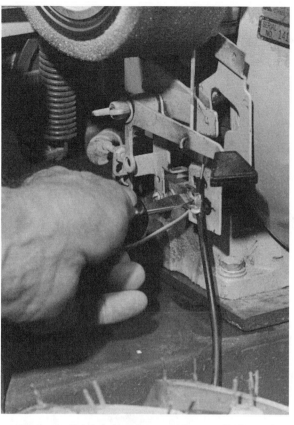

Soak a little light oil or an oil/kerosene mix into Bowden cables before storage, as shown above, to prevent rusting and sticking. Also be sure the small clamp that secures the cable is snug.

stop switch and choke action don't relate to the markings on the quadrant.

Change the engine oil. When changing oil you should accomplish two things. You should remove the old oil, which may have become corrosive from the effects of minute amounts of sulfur in the gasoline mixing with the water that is always produced by combustion. You should then circulate clean fresh oil to coat all of the vulnerable parts. This means running the engine until the old oil is thoroughly hot and then draining it promptly, followed by a second run with new oil and (if your engine has one) oil filter.

The second run is a good time to run the carburetor dry of old fuel, which would otherwise leave a tenacious, gummy residue that plugs carburetor jets and sticks float needles and diaphragms. That will also remove most of the gasoline from the fuel lines and pump. Before running, however, you must remove all of the old fuel from the tank.

Drain the gas tank. On some engines this can be done by lifting the tank free from its retainer clips and fuel line and inverting the tank. Other tanks may have a drain plug, and with some it's easiest to disconnect the fuel line from the bottom of the tank or the carburetor inlet. With a light rotary mower it may be simplest to flip the entire mower over and let the old fuel drain into a large, shallow pan. You'll find this easiest after the old oil is drained and just before adding the new oil in preparation for the second, shut down run.

Here is one operation where the usual precaution of removing the wire from the spark plug doesn't apply. If an accidental turn of the crankshaft generates a spark, it's better to have the spark inside the engine than jumping from a loose plug wire when

Some engines have a convenient removable gas tank that slips out of retaining clips. Be sure the fuel hose is secured at the carburetor on reassembly.

After running the tank dry, remove the last dribble of fuel by pushing up on the drain valve beneath the carburetor bowl.

draining gas. *Do all such draining well away from your home, garage, or shed, do not smoke, and always provide a suitable catch pan for the drained gas.*

Clean and inspect the engine exterior. Just before the shutdown run is also the time to remove oil, gum, and dirt from the outside of your engine. Use a brush and kerosene or mineral-spirits paint thinner. Be wary of spray-on automotive engine cleaners; they can be death to some plastic parts and fuel hoses. Use carb cleaner sparingly around the throttle and choke shafts. Do this just before running so that any cleaner residue will evaporate or be pulled through the engine.

Protect the fuel system. Rather than simply run the carburetor dry of old gasoline, a better bet is to mix 1 ounce of protectant fuel conditioner with a gallon of fresh gas and run on this mix until the carburetor is dry and the engine stops. The inside of the tank, fuel lines, and carb will be protected against gumming. You'll find the product at most outboard motor dealers and other marine supply shops.

Protect the cylinder wall. Most marine-supply shops also sell pressurized containers of storage fogging oil to spray into the carburetor throat during the shutdown run. Fogging helps keep the cylinder wall and other parts from rusting due to moisture in the air that condenses on the cold metal surfaces. Wait until the carburetor starts to run dry, and just before the engine stops introduce a quick shot of fogging oil. This will also protect your engine in case a quick shutdown has caused raw gas to wash the oil from the cylinder wall.

In lieu of fogging through the carburetor, you could remove the spark plug after the second run and squirt light oil into the cylinder. Follow this by pulling the starter rope or operating the starter briefly with the plug still out. Do not replace the plug before cranking, since you may have introduced enough oil to block the piston and bend a connecting rod.

Close the valves. Your engine should now be well protected internally. It's also good practice to turn the engine by hand until you feel compression. That means both valves are closed and should help seal against moisture entering. On some engines with

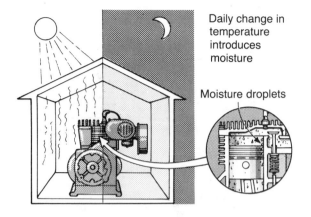

Daily change in temperature introduces moisture

Moisture droplets

A typical garden shed gets hot under the sun's rays. At night as the engine chills down, moisture in the air condenses on the cylinder walls as shown above. Fogging through the carburetor (below) helps prevent rust and corrosion.

gency generators that may be called upon at any time to power your house, however, there is no good alternative to starting and running under load at least once and preferably twice a month. This not only helps prevent gumming and moisture accumulation in the engine and ignition and electrical parts, it also keeps the generator windings dry and retains the residual magnetic charge in the generator field so current will build up.

Use protective fuel mix. If you maintain a constant supply of fuel in your generator tank so it's always ready to start, use the 1 ounce of fuel conditioner per gallon of gas previously described. The conditioner should stabilize the fuel for at least one year. When shutting down always run the carburetor dry.

STORING TWO-STROKE ENGINES

With no oil in the crankcase and lubrication supplied by a constant mix of fresh oil and fuel, you do not have the same concerns about dirty, acidic oil

automatic compression release, it may be difficult to feel compression with the plug in place. Try feeling for it with your thumb over the plug hole before replacing the plug.

Storing Emergency Generators

Small portable generators used for camping trips and the like should be stored the same way as other small four-stroke engines. For engine-driven emer-

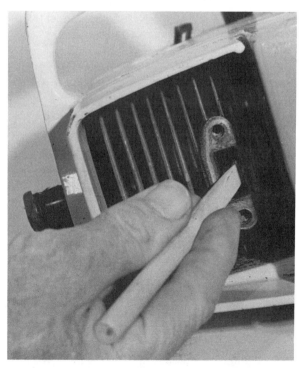

A hardwood stick or old toothbrush handle sharpened to a chisel edge works well for poking out carbon from a two-stroke's exhaust ports.

and draining as you do with a four-stroke. It is still good practice to fog the carburetor while running the fuel system dry on a mix of gasoline, oil, and fuel stabilizer. External linkages should also be cleaned up and residual dirt, such as sawdust on a chainsaw, should be wiped or blown away.

Remove the carbon. There is one routine maintenance operation on a two-stroke that is not required on a four-stroke, and that's removing carbon from the exhaust ports and muffler. Using a high-quality two-stroke oil may leave the ports practically clean, but there is no way of knowing without removing the muffler or spark arrester and inspecting them.

If a buildup of black carbon is visible and partially clogging the ports, turn the crankshaft until the

For a handy plug-grounding tool (left), break the top off an old spark plug, insert a length of brass wire through the center, and solder the wire at the terminal. Below: Blade is removed for seal access with a suitable puller as shown here.

One wipe with your finger tells you the lower seal on this engine is leaking.

ports are blocked by the piston. Then remove the carbon *as gently as possible* with a tool such as a hardwood dowel or piece of toothbrush handle sharpened to a chisel-like end. Do not use a screwdriver or any tool that might scratch the piston.

Inspect the seals. A two-stroke engine depends upon crankcase compression to operate. This compression is retained by the crankshaft bearing seals. A common problem on vertical-shaft mower engines is the tendency for the spinning shaft to pick up fish line, kite string, and similar materials and wind them tightly around the shaft in the seal area—where they often cannot be removed without damaging the seal.

To check crankshaft bearing seals, tip the deck up to a vertical position, exposing the mower blade. Remove the spark-plug wire from the plug and ground it to the engine to prevent accidental starting while pulling the blade.

Plastic garbage bag protects this snowblower engine from rain and snow. Wait until the engine cools before slipping it into place.

Many pieces of yard equipment have synthetic rubber fuel lines or fuel-line sections. Age and heat deteriorates such line, and if it leaks there's a fire hazard. Check it at least once a year.

This ignition switch can fill with water on machinery stored outside. A bit of tape over its face saves lots of hard-to-trace trouble.

With the mower blade removed, you may be able to see the lower shaft and oil seal area with a flashlight, or you may have to remove the blade-mount hub and dirt shield. If the hub is secured by a bolt into the end of the crankshaft, it may be best to loosen the bolt before removing the blade. That way you can block the blade to keep the shaft from turning. Otherwise, you'll probably have to block the piston by inserting either a blocking tool or a length of sash cord through the spark-plug hole.

You'll probably need a puller to remove the hub. This can be a problem since the hub fits over the end of the shaft and leaves you with nothing for the puller screw to push against. The trick here is to use a bolt the same size and thread as the original blade-retainer bolt, center-punch the bolt head, and drill a small depression in it to center the puller screw. Run this bolt partway into the hub and pull against it. The hub will come only as far as the bolt head projects, so you may have to back the bolt out as you continue pulling.

Blade brake/clutches don't make blade-mount hub removal any easier. Before attacking one of these, consider the possibility of removing the engine-mount bolts and lifting up the engine a bit for a peek underneath. Look for two things: material wrapped around the shaft and/or a significant buildup of oil. In either case, the seals are in trouble and must be replaced. Considerable care and a few simple tools are needed, as outlined in Chapter 15.

Protect the reed valves. The reed valves are the second vulnerable point on two-stroke engines that have them. You may be able to see the reed valves by looking in through the carburetor with a flashlight. If there are none, you'll see the side of the piston. If your engine has reed valves, tip the engine up so the carburetor is vertical and squirt in some light oil to help protect the reeds from rusting during storage. You can also do this by fogging the engine through the carburetor.

Protective Coverings

The most direct action is simply throwing a sturdy box over the engine and perhaps weighting it with a piece of firewood. A better plan is to use a canvas bag or cover that can be drawn snugly around the engine. A military surplus duffle bag usually works fine. For a quick job, try several heavy plastic trash bags, one inside the other. Just don't put them over a hot engine or they'll melt and stick.

Preparing A Battery For Storage

1. Add distilled water with a squeeze-bulb filler to bring acid level up to marker. Avoid getting acid in eyes or on skin or clothing. Use safety goggles.

2. Brush baking soda and water onto whitish acid deposits to neutralize them. Do this with the covers on to keep the mixture from getting inside.

PROTECTING NON-ENGINE PARTS

In spite of all the careful measures you take to prevent storage damage, there is still a good chance that something not really a part of the engine will cause trouble when you try to start up next season. These are often parts supplied by the machine manufacturer rather than the engine builder.

Inspect remote tanks and lines. On larger tractors where the fuel tank is at the end opposite from the engine, remove the tank filler cap and examine the interior with a flashlight for rust and silt. Look inside at the tank outlet for a small strainer or screen. There should also be a fuel shutoff valve under the tank at this location. If dirt is present, remove the

3. When the deposits have bubbled and fizzed for a minute or so under the solution, thoroughly flush the battery case and surrounding area with a hose.

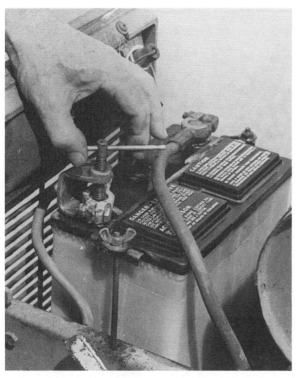

4. If the cable terminal does not release with gentle twisting after loosening the clamp nut, use a terminal puller to free it. Never hammer or pry.

5. Often, simply cleaning terminal posts will bring a "dead" battery back to life. Use a cleaning tool as shown or medium-grit sandpaper.

strainer and valve. Hold the strainer against the light and if it appears gummy, spray it with carburetor cleaner and blow it dry.

Check the fuel line. Examine every inch of the fuel line from the tank to the engine. Look for sections of synthetic rubber line that may be cracked or worn. Also look over the metal lines for wear spots from vibration or rubbing on the chassis.

Unless you can shut off the fuel at the tank, it is hazardous to store a tractor, mower, or the like in your garage or shed. Even a slow leak in a line can cause a buildup of explosive fumes. And if a line ruptures while you are operating the machine, you may find yourself soaked with gasoline or on fire.

Protect the switches. Cover electrical switches on

6. Don't forget to clean the insides of the cable terminals. Again, the tapered reamer on a suitable cleaning tool works best, but rolled sandpaper will do.

machines stored outdoors where rain or snow could enter them. Use weather-proof tape. If you can get to the rear of the switch and the connections, coat these areas with silicone sealant to prevent water entry. Also protect safety interlock switches under the seat and those involved with the brake, clutch, or implement engagement. Any of these can prevent engine starting if shorted out by water or corrosion.

Storing Batteries

If you have an electric-start, walk-behind mower, check the manufacturer's instructions for battery storage. Add water to lead/acid types as needed, and bring to full charge either by running the engine or with a charger. Remove the battery and store in a dry, reasonably warm place away from any source of flame. If the battery has deposits on the outside and terminals, wash with a baking soda/water solu-

tion and rinse clean. Clean terminal posts and terminals to bright condition on the contact surfaces.

TAKING YOUR ENGINE OUT OF STORAGE

When the new season rolls around, if you properly mothballed your engine, start-up should be a pleasure. If you did not, you may need to check all of the previously mentioned points.

Begin with a cleanup. Presuming that you did store your engine well, you obviously want to remove any storage tape, heavy grease, or other protective materials. The second step is to make sure that nothing unexpected has intruded during storage. It's not unusual to find a mouse nest, bird seed, or similar messes inside a blower housing or air

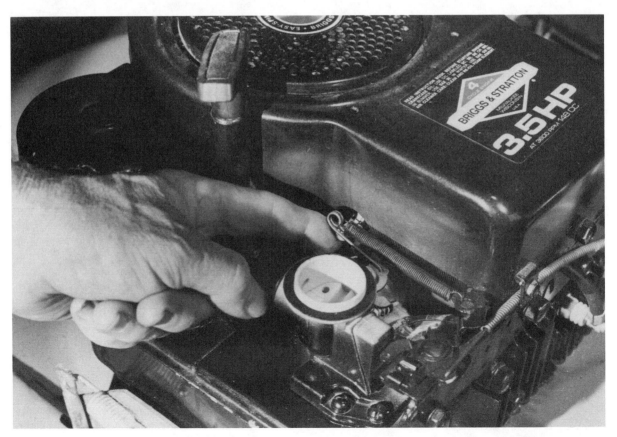

A quick, gentle poke on linkage and spring parts will tell you if all are properly connected and working nicely.

Some gasoline inevitably works its way out around the throttle shaft and choke, making them stick. Carb-cleaning spray will quickly clean away the gum left behind.

cleaner, or even in an exhaust pipe. Mice will also gnaw electrical wires bare.

Check the throttle and choke. Remove the air cleaner and operate the choke control if the engine has one. Make sure the control operates freely and that the choke opens and closes completely. Do the same with the throttle. The throttle will probably be linked to the governor arm or spring. If the control and the governor linkage are operating properly, moving the throttle control to wide open should open the throttle fully. Feel for free movement of the throttle butterfly and linkage parts. It's disconcerting and dangerous to have your engine start and run wildly because the governor is sticking or the throttle hanging wide open.

Test the spark. If you fogged oil through the carburetor, you may have fouled the spark plug. You'll want to check it anyway, so remove the plug and look for traces of oil. A quick cleanup can be made by squirting a little carburetor cleaner or starting fluid on the electrodes.

If the plug is clean, new, and properly gapped, put the controls in the start position with the plug grounded to the engine. Pull the starter rope if the engine is manual start. The engine should spin freely with the plug out and you should see a healthy spark. If you hear or feel anything that suggests binding of the internal parts, now is the time to check for rust.

Recheck the oil. Even though you filled the engine

with fresh oil at the end of the season, it's always possible that a leaking seal, gasket, or plug allowed some or all to run out. Sometimes, if your engine has been covered with snow or rained on, you'll find water in the oil.

Look for gas leaks. For safety's sake, put only a small amount of gas in the tank. Note whether gas leaks around the fuel valve stem. A little tweak on the packing nut might be needed if the packing has dried out. With the valve open, run your fingers along the line to the carburetor or fuel pump and carburetor. If you find no leaks, fill the tank and your engine should now be ready to go to work for the upcoming season.

TROUBLESHOOTING AND REPAIR

Troubleshooting Starting Problems

Balkiness was once almost a built-in feature of small engines. This was back when you wound a rope around the flywheel pulley and tugged mightily. Today's little engines are so good that my rule is, if it doesn't start on the third pull of the starter you've probably got a problem. The first two pulls may be needed just to get some gas into the carburetor, but after that the engine should fire. On some small engines with pump-type carburetors such as on chainsaws and weed trimmers, five or six pulls might be needed but that's all.

If you were reasonably careful when storing your engine, it will probably start readily at the beginning of the season. Assume, for the purpose of this chapter, that your engine has been running previously but won't start now.

GENERAL STARTING TECHNIQUES

Some engine owners seem to lack any rapport with their engines. Those who were driving before cars had automatic chokes have probably developed an intuition as to how an engine is responding. One can more or less learn how to nurse an engine to life but that shouldn't be a constant necessity.

Starting a four-stroke. Starting a cold four-stroke engine is sometimes easier than a two-stroke, although the heavy oil may make manual starting feel heavy and difficult. To pull the starter on a 10-hp snowblower in freezing weather, for example, you must brace your feet firmly and exert considerable force. If you have a heart problem or other physical limitations, I strongly recommend an electric-start kit if one is available for your engine.

Begin with the choke closed. Some four-stroke engines will pop and stutter like a two-stroke, a sign that the choke should be opened partway to lean out the fuel mixture with more air. Others will start and run sluggishly, accompanied by black smoke—signi-

fying that the mixture is still too rich with fuel. Open the choke gradually until the engine smooths out and the black exhaust smoke clears. As with old cars, if the engine starts to die give it a bit more choke, and if it bumbles along ease off the choke.

Starting a two-stroke. Although some two-stroke engines have primers and so-called automatic

If your chainsaw has gone unused for some time, it may take five or six pulls with the choke closed before those first pops signal that fuel is getting to the carburetor. Be sure to hold down the saw securely with either your knee or foot.

chokes, it still helps to know the typical sounds of over-rich and over-lean conditions. With most small two-stroke engines such as chainsaws, the conventional starting technique calls for closing the choke when the engine is cold. With the ignition on, two or three pulls should produce a muted pop or stutter, which means fuel is reaching the combustion chamber.

Continuing to pull the starter with the choke closed will almost always result in flooding and no start. Listen for the first sounds of combustion and then open the choke. If a wide-open choke produces a brief run during which the engine seems to grow weaker and die, repeat the sequence but this time open the choke only about halfway. The engine should start and as it gradually warms up you can ease the choke open further until it runs smoothly

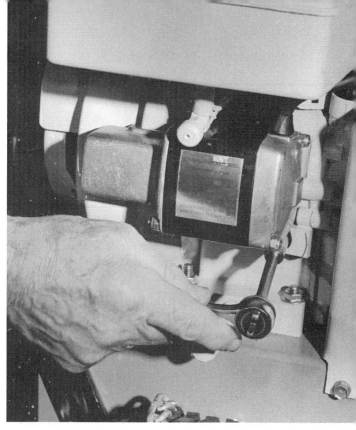

Snugging down on the mounting bolts is the final step in equipping this snowblower engine with an optional electric starter kit.

with the choke fully open. On some engines, such as a snowthrower in cold weather, you may have to work with a partially closed choke for several minutes until warm.

Starting a hot engine. As a matter of normal practice, try starting a hot engine *without* the choke. If it doesn't start with one or two pulls, try partially closing the choke. Some engines, especially those with diaphragm carburetors mounted on top of the tank, seem to require full-choke starts to establish fuel delivery even when warm or hot. In other cases, shutting down a hot engine for a few minutes may cause the fuel in the line and carburetor to vaporize from engine heat. In effect, the fuel boils out of the carburetor or creates a vapor barrier in the line—a condition called *vapor lock.* Again, a choke usually helps.

Most small engines have rather individual starting characteristics. What works well for one may not be suited to another. The trick is to listen and learn to identify the various sounds and the feel of an engine during starting.

If your two-stroke engine is flooded, try removing the spark plug and pulling the rope rapidly. Pops and poofs of flame through the plug hole are sure evidence of excessive fuel. Do this with the switch on and the plug gap near the plug hole.

Starting a flooded engine. A badly flooded two-stroke engine often sounds "wet," and what you're hearing is excess liquid fuel in the crankcase and combustion chamber. The quickest cure is to remove the spark plug and pull the starter several times. Dry off the plug if it's wet. One sure sign of flooding is a "bloop" of blue flame when you pull the starter with the plug grounded next to the plug hole. That's excess fuel being expelled. If you replace the plug after the blue flame subsides, you may be able to start the engine or you may have to wait until more fuel evaporates.

Cold-Weather Starting

Hand-starting a four-stroke engine of 5 hp or more when the temperature is 15°F will usually be much harder than starting that same engine on a warm summer day. Even with an electric starter, your small engine may be much more stubborn than the one in your passenger car. Stale or summer-grade gas, low cranking speed, or drag from something like a belt-tightener clutch or hydrostatic-drive transmission could be to blame. The main factor, however, is that you're only working with the power impulses from a single cylinder. If the first power stroke doesn't produce enough torque to rotate a four-stroke engine through one and a half more revolutions against internal friction and drag loads, the engine may "pop" but will not start.

This is especially true for hydrostatic-drive garden and lawn tractors if the drive cannot be disengaged. Even though the drive is in neutral, the engine will fire repeatedly but won't start.

Use thinner oil. Tecumseh, for example, recommends SAE 30 oil above 32°F for their four-stroke

A little heat can be an enormous help in starting a stone-cold engine. This soldering iron is secured to the side of the crankcase with baling wire. Pack some glass-wool insulation over it to confine the heat.

engines; below 32° they recommend 5W20 or 5W30, and below 0° they suggest 10W diluted with 10% kerosene. The heavy oil should be drained while hot, the thinner oil put in, and the engine operated briefly before cold weather. Trying to drain cold SAE 30 oil at sub-freezing temperatures is tedious and still leaves the thick oil on the piston and bearings.

Apply heat. Many engines have long dipstick tubes. If you can slide a dipstick heater down into the oil a few inches you should have no trouble starting. With a hydrostatic tractor-drive, you'll find a dipstick heater does far more good if you put it in the transmission oil-check hole rather than in the engine oil.

If you don't have a dipstick heater, simply place a heavy-duty soldering iron along the engine base so both the oil and the carburetor receive some heat.

Check the battery connections. Battery starting capacity drops sharply in the cold, so it's important that the terminal connections are bright and secure. It is also a good idea to keep a trickle charger on the battery for at least a few hours before starting. The electro-chemical action will warm the battery and greatly boost its cranking power.

A tiny burst of starting fluid into the carburetor and a quick pull on the starter will usually tell you if you have ignition and compression.

Use starting fluid. I have several otherwise-reliable engines that simply won't fire in cold weather. To start, I routinely remove the air-cleaner cover, open the choke, and shoot in a small burst of starting fluid. Then I close the choke, hit the starter button, and away I go.

STARTING PROBLEMS—IGNITION, FUEL SYSTEM, AND COMPRESSION

Nearly all service manuals and instruction books carry neatly presented troubleshooting guides. Whether these are lists of hypothetical troubles and possible causes, or elaborate tree charts you must trace step by step to find the trouble, I've seldom found such guides very helpful. Often the small-engine owner works for hours following false leads, probably doing a lot of unnecessary work and disturbing parts best left alone.

When the owner finally takes his engine into a repair shop, the mechanic has it ticking in a few minutes. That's because the mechanic did not spend his time reading a list and looking for unlikely trou-

A wooden clothespin makes a good insulating handle when checking for a spark.

ble sources. What he did do was reflect in his own mind, consciously or intuitively, on the basics of what makes an engine run and start checking these first. The first questions to ask yourself are:

- Do I have ignition?
- Is fuel getting to the combustion chamber?
- Do I have compression?

Later, after the pro gets the engine running, he may discover other problems such as missing, smoking, erratic speed control, and so on. These details are a tipoff to other repairs that may be needed, but first he'll determine if he has a runnable engine.

Checking the ignition. Probably the first check a pro would make is the easiest—namely, removing the spark plug, reconnecting it to its lead wire, then placing it on top of the cylinder head and pulling the starter. This is also your first step. If you see a healthy, bright-blue spark, you probably have ignition; if not you'll have to dig into that, but for now assume you have a good spark.

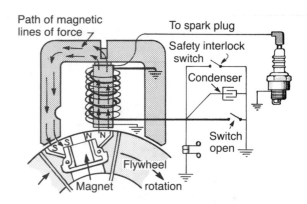

Magneto system, above, generates its own electrical power from the interaction of the flywheel magnets and primary coil windings. In a battery system (below), power is supplied by the battery.

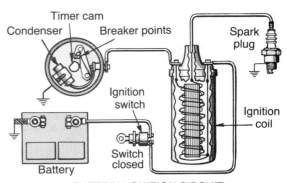

BATTERY IGNITION CIRCUIT

Checking the fuel system. If you doubt that fuel is reaching the combustion chamber, try a brief shot of starting fluid into the carburetor throat with both choke and throttle open. Crank the engine immediately since the fluid evaporates fast. If you get a short, fast run on the fluid you know you are getting ignition and probably have adequate compression but are not getting fuel from the carburetor.

At this stage, a second trick may work if the carburetor jets are slightly plugged with gum or oil. Repeat the above start with the fluid but have the choke closed or snap it closed during the brief run. In many starting problems with engines out of storage, this was all that was needed to clean out the jets for the rest of the season. Once the engine was running, the flow of fuel flushed the jets clean. *If you try starting fluid on a chainsaw, be sure you have the saw held down firmly.* Sometimes the engine will kick

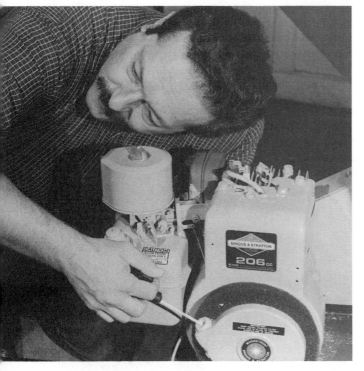

A wheezing sound at the muffler when you pull the piston up on compression is a sure indication that the exhaust valve needs servicing.

How To Read A Spark Plug

A spark plug can't fire the fuel if its gap is bridged by combustion deposits. The cause may be oil additives, bits of loose carbon, or even foreign material that bypassed the air filter.

This plug has been running too hot, either because it's the wrong heat range, the mixture was too lean, ignition timing was off, or the engine was overloaded. Note the very white central core.

Preignition damage to this plug suggests some of the same conditions that cause running too hot. In such cases, inspect the piston for similar damage. You may find that it has a hole in it.

Wet deposits such as this are commonly found in two-stroke engines. The wrong oil or fuel/oil mixture is probably the cause. Such a plug on a four-stroke engine usually denotes worn valves or piston rings.

This plug is simply worn out and should be replaced. Note the severe electrode erosion and deterioration of the center core.

Carbon deposits can result from the wrong heat range, too rich a mixture, or a weak ignition system. Basically, the carbon is fuel residue from incomplete combustion.

Above photos courtesy of Champion Spark Plug Co.

and pull the rope hard enough to spin the saw towards you.

Checking compression. Assume for the moment that the starting fluid did not help. You observed a spark but providing fuel to it did not produce even a brief run. Begin your compression check by removing the spark-plug wire and rotating the flywheel with your fingers or with the manual starter. You should be able to feel whether the piston is building up compression or not. If the flywheel spins easily through two revolutions with little resistance or bounce, you can suspect a blown head gasket, stuck piston rings, or a stuck or burned valve.

TROUBLESHOOTING THE IGNITION SYSTEM

Although Chapter Ten provides many more details about ignition systems and their servicing, there are a few basics you should keep in mind just for get-it-started troubleshooting.

Most small gas engines use *magneto ignition.*

Whether the system is solid-state or traditional breaker-point, this means that magnets are built into the flywheel rim and the engine generates its own electricity without a battery or other external power source. A few engines, particularly Kohler, do use battery ignition, but these are mainly industrial applications.

Ignition switches. The first rule to remember is that all magneto, non-battery systems generate their own power. For the engine to run, the ignition switch and all interlock and safety switches connected to the system must be *open.* Close the switch and you ground out the whole system and the engine won't start.

This is exactly the opposite of battery-powered systems such as the one on your car. With battery power the switch must be *closed* to feed battery power to the circuit.

Spark-Plug Service

Many small-engine starting problems relate to the spark plug. The plug may be fouled, internally conductive, or the wrong heat range. The spark plug is always the first thing to check when starting problems arise.

Check for fouling. Dampness from outside storage, as well as dirt and grease from fingerprints, may cause a plug to misfire. Also, condensation inside the cylinder or intake-valve port in a four-stroke engine can cause a drop of water to be splashed onto the plug electrodes and short out the plug. A splatter of oil or minute fragment of carbon or other combustion debris can do the same thing, particularly in two-strokes.

If the plug looks oily and the engine is a four-stroke, it suggests that the piston rings and/or the valve guides are worn. On a two-stroke engine a little oil is not unusual as long as it's just a little. Most four-stroke engines will show light gray or reddish-tan deposits on the spark plug core nose. If these are heavily deposited in the space between the core nose and the outer shell, it's not really worthwhile to try to pick them out. In fact, cleaning spark plugs is generally a waste of time, and sandblasting is definitely bad practice. There's just too much chance of sand wedging up in the plug and then dropping out and damaging the cylinder and piston. If you've logged over 50 hours on a four-stroke spark plug—or over 20 on a two-stroke—replace the plug.

Many deposits around the core nose are conductive and can provide a path to ground for the high-

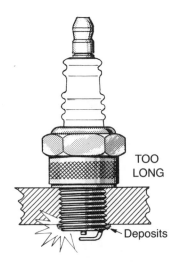

When you try to remove a plug that's too long, the combustion deposits on the lower threads can cause hard turning and may destroy the threads in an aluminum cylinder head.

voltage energy, again, shorting out the plug gap.

Spark-plug reach. The length of the threaded portion of the plug is called its *reach.* Some chainsaw engines and the like may require plugs with only a ¼-inch reach. Others may require as long as ⅝-inch, and there are several lengths in between.

If you install a plug that is too long, the piston may strike it and cause severe damage. A more likely condition is a plug that projects only a thread or two past the inner face of the cylinder head. Under the heat of combustion, such projecting plug threads or deposits on them can glow, and ignite the fuel prematurely—a condition called *preignition.* Besides reducing power preignition often leads to *detonation,* or the uncontrolled burning of fuel commonly called pinging or knocking. The costly consequence, in small engines with light aluminum pistons, is often a hole in the piston crown or a burned-off piston edge.

The same damage can occur if the plug reach is too short. Now the uncovered threads in the cylinder head become hot and glowing, again causing preignition and detonation.

Both situations also cause trouble when removing or installing a new plug. Deposits or corrosion on an extra-long plug will prevent free turning in the threads, making it extremely hard to remove. Such deposits can literally tear the threads out of an alu-

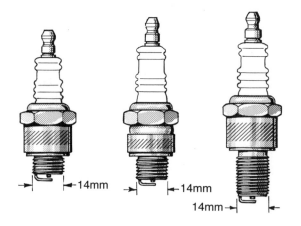

Even though all three of these plugs are 14mm, the wrong length or *reach* could cause serious trouble from detonation or even interference with the piston. Always use the plug reach specified for your engine.

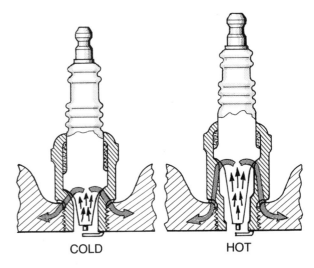

COLD   HOT

When you replace a short plug with one of the proper length, you'll probably have to use a tap or plug-thread cleaner to remove deposits that would prevent the new plug from seating.

Spark-plug heat range is controlled by varying the length of the path the heat must travel through the core to reach the cylinder head, through which the heat is carried away.

minum head if you force the plug. Instead, try to get some penetrating oil down into the threads and gently work the plug back and forth with a socket wrench.

If the old plug was too short, the deposits in the last few threads will prevent the new plug from seating against the gasket properly. Run a tap or cleanup tool all the way through until the threads are formed cleanly.

Spark-plug heat range. You'll hear plugs referred to as "hot," "cold," and sometimes "medium" plugs. That has nothing to do with the intensity of the spark delivered, which remains the same. It does relate to the temperature at which the plug-core nose operates. If the plug core and electrodes run too hot they will erode or burn away quickly and may cause detonation. If these parts run too cold they will collect deposits and foul. A plug of the proper heat range will more or less self-clean without rapid erosion.

Sometimes an engine manufacturer may have anticipated a heavier load on the engine than the job you're doing and specified a cool plug. If your plug rapidly builds up deposits and causes starting problems, try the next-hotter plug. Conversely, if your plug shows white blistering and rapid erosion of the electrodes, try a slightly cooler plug.

Anti-RFI plugs. In many areas, particularly in

Canada, stringent laws prohibit equipment that produces radio frequency interference. In most small engines RFI is suppressed by resistor-type spark plugs. If the designation code on your original plug

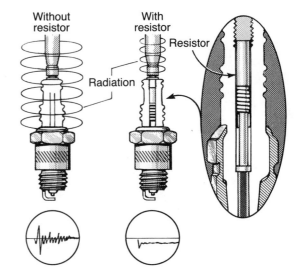

Spark plugs and ignition wires cause radio and TV interference by radiating high-frequency energy waves. A resistor inside the plug core greatly reduces this radiation without affecting the plug's performance.

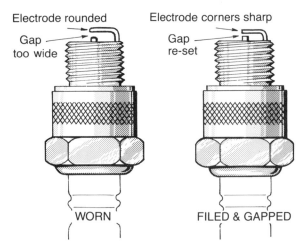

Electrode rounded   Electrode corners sharp

Gap too wide   Gap re-set

WORN   FILED & GAPPED

You can improve a worn plug by using a fine file to sharpen the electrode corners and faces. If the plug is oily, wash it with lacquer thinner or the like and then dry it off.

includes the letter R, be sure the one on your replacement plug does, too.

Removing stubborn spark plugs. Even plugs of the proper heat range and reach can sometimes be unusually difficult to remove. If the plug and gasket are gas-tight, it's unlikely that penetrating oil will get down into the threads. The only real choice is to use a good socket that fits the plug, a sturdy flex-handle, and a strong pull. A ½-inch-drive wrench may be necessary. Just barely break the plug loose. Then, if it doesn't come freely, the penetrating oil may help because it can now get past the gasket and down through the threads.

Baked and hardened fuel gum in the threads can also make them stick. If the plug wasn't quite tight, combustion gases may have worked up along the threads and literally glued the plug in place. For that reason, it doesn't hurt to add a little carburetor cleaner to the penetrating oil while it's soaking. Even kerosene will help if you gently work the plug back and forth.

Gapping the spark plug. The gap between the spark-plug electrodes should be checked both before installing a new plug and when you've removed a used plug that appears to be good. Consult your manual for the correct gap, and adjust if needed with a wire-type feeler gauge/bending tool. Do not force the gauge through; it should have only a barely perceptible drag.

With a used plug, the gap has usually opened somewhat and the electrodes may have lost their sharp corners and edges. Unless the outer electrode is so worn as to show a notch, use the bending tool on the feeler gauge to carefully move it away enough

Take a hard look at the ignition wire where it exits the blower housing. Chafing or wear can allow the spark to ground out at this point.

If the small switch-grounding wire shown here touches metal or is shorted by moisture or dirt, your engine won't run.

to introduce a thin, flat ignition file. You can also bend the outer electrode with the tip of a small screwdriver levered against the metal outer shell. *Never try to move or bend the center electrode.* Instead, file its tip so that it is flat and the edges are sharp. Also file the tip of the outer electrode to produce a sharp, square edge. It takes less ignition energy for the spark to jump between two sharp edges than between two worn, blunt surfaces.

You should be able to run the plug in with your fingers or very light wrench action until it seats on the gasket. On new plugs, standard recommendation for final tightening is to pull the wrench handle one-quarter turn after the plug seats firmly on the gasket. On used plugs where the gasket has already been compressed, pull just hard enough to feel the plug turn firmly against the gasket. In rare applications with no gasket and a tapered seat in the head, one-sixth turn is recommended.

Non-Plug Ignition Troubles

Once you know you have a good spark plug, yet still don't have a spark or one hot enough to start your engine, check the high-tension lead wire to the plug at the point where it emerges through or from under the blower housing. Gently tug the wire out about ¾ inch and examine it all around. Look for chafing, cracking, or abrasion, which may be letting ignition energy leak away. In such cases, testing the plug for spark outside the engine may indicate no problem, but inside, the greater resistance across the gap caused by the compressed fuel charge may be making the plug misfire. If you get a spark with the plug out, reinstall it and, with the wire still pulled clear, try starting the engine. You may be surprised to see it fire because the energy leaking out of the lead wire has found its way to the plug.

Nearly all high-voltage leads are now molded or sealed into the coil, and you might need a new coil and wire assembly. But first, try coating the chafed spot with silicone sealer. Don't use the metallic variety. Let it cure overnight and it will probably last as long as the engine.

Primary wire problems. On many mower and tiller

If you suspect that a grounded switch is preventing ignition, disconnect the ground wire and isolate it with tape or a wooden clothespin. If you get a spark, you've found your problem.

Typical breaker points, coil, and condenser are found beneath the flywheel. The only real check for a defective coil requires a coil tester.

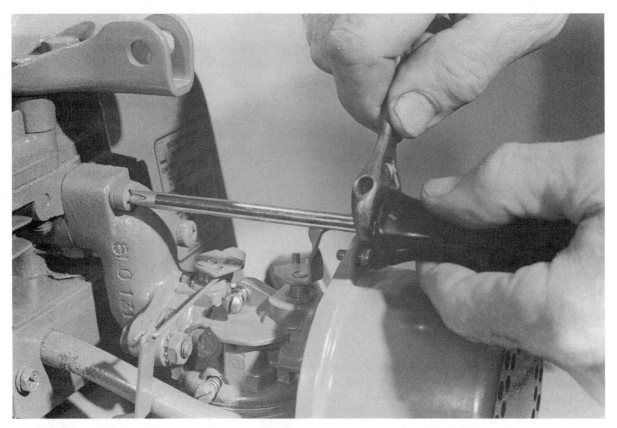

Sometimes a loose carburetor flange or intake fastener will cause an engine to drop off in power and then surge back as though the fuel flow were erratic. Snug up such fasteners.

engines the remote throttle control is marked *Start, Run, Slow,* and *Stop.* Move the control to stop and the throttle linkage slides a little switch closed and shorts out the ignition system. Such a switch will probably be on or near the carburetor and will have a light-gauge wire running to it. This wire is usually just clipped to the switch and is easily pulled or broken off accidentally. This seemingly innocent wire is the magneto primary-ground wire, and if it's touching bare metal your engine will not start. Sometimes just mowing or working under low shrubs will snag it.

If your engine won't start with the wire in place, try slipping it free of the clip on the switch. Wrap a fold of tape over the bare end and see if the engine starts. If it does, the switch is defective or fouled. Clean it with a small brush and solvent and examine it carefully. You may find that the sliding member is bent, damaged, or corroded. Plan to get a replacement. Meanwhile, you can start and operate with the ground wire taped. Stop the engine with the choke or by shorting the plug with a screwdriver if it has an open terminal.

Chainsaw switches. On other types of equipment such as chainsaws, a small manual switch stops the engine. Sometimes you can get at the terminals of these switches easily, and other times you have to disassemble half the machine to open up the handle. Unless you're prepared to open up a tight package of controls with many small parts, take the saw to a dealer.

Remote switches and interlocks. On some equipment it's easy to spot a low-tension wire running up to a mower or tiller handle or under the seat of a rider mower or tractor. These are also magneto-ground switches. If the switch itself is buried, it may be easier to snip the wire at a handy point and see if you get a spark. Later, with a new switch or other trouble source revealed, you can solder and tape the wire back together.

In addition to the main switch and the underseat safety switch on many machines, there may be several other interlocks connected to the ignition ground and possibly the starter circuit. These may be somewhere in the implement drive clutch to prevent starting with the cutter engaged, and/or near the shift or brake pedal/clutch to prevent starting except in neutral with the foot brake depressed. Usually, these switches are connected with slip-on terminals. Simply disconnecting these to be sure they aren't touching metal will enable you to check for ignition ground.

Don't overlook the point that although the switch may be in perfectly good shape, the wire leading to it may have been chafed or pinched in the machine chassis and be grounding out there. While you're checking switches and wires, also take note of other small wires and connections and try them for tightness.

Inside the ignition system. When you've done all you can with the external parts of the ignition system, the trouble is probably somewhere in the magneto itself. It could be a corroded or worn set of breaker points if your engine has them. It could also be a bad condenser or coil, or a bad electronic ignition module. Chapter Ten covers in detail how these parts operate and how to service them.

By now, unless your ignition system is really out of business, you should be able to start—in three pulls or less.

TROUBLESHOOTING THE FUEL SYSTEM

Begin by taking a quick look at the carburetor with the air cleaner off, and be certain that the governor action and throttle controls have really opened the throttle wide after you open the control lever. If there's a choke plate or other choke device be sure it's closing properly. And—by no means as silly as it seems—try shaking the carburetor to be sure the mounting nuts or screws are secure. Many hours have been wasted trying to start or adjust a carburetor when all that's wrong is that it is loose and leaking air at the mounting flange.

Also try closing the choke when turning the engine. You should be able to see raw fuel in the carburetor throat. If there is fuel, although the carburetor may be misadjusted or have some other trouble, the engine should at least pop or backfire.

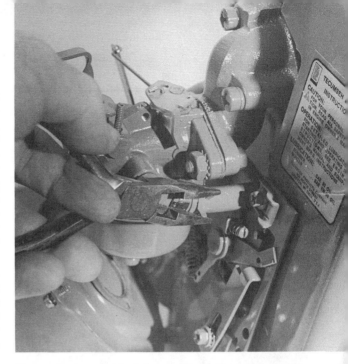

Disconnecting the fuel line at the carburetor should show a free flow. Do this outside and don't smoke while you do it.

Fuel shutoff valves can get sticky and even plug. If you can't turn yours with your fingers, remove it and clean with carburetor solvent. Be prepared to catch the fuel that runs out.

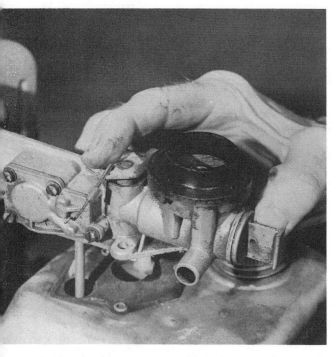

If the intake screen on a tank-top carburetor is gummed or dirty, you'll get no fuel and no start. You'll have to lift the carburetor free of the tank to check.

Instead of the usual flange and gasket, this carburetor has an O-ring seal that slides over the intake. Be sure this seal is not damaged or worn out.

Check tank-to-carburetor connections. If you get only short runs of a few seconds after using starting fluid, you can be almost certain of one of two conditions, and maybe both. First, the fuel may not be getting from the fuel tank to the carburetor. If a rubber or metal line runs from the tank to the carburetor, as it does on a typical gravity-feed system, loosen the line connection at the carburetor and see if fuel flows. If not, the line could be plugged, but more likely a small screen or strainer in the tank is blocked with fuel gum. Drain the tank out in the open away from the house, providing a receptacle for the gas. Then remove the strainer—usually together with a shutoff valve—from the bottom of the tank. Usually all that's needed is a good washing and brushing with lacquer thinner or carburetor cleaner. Clean until you can see through the screen and blow air through it and the valve freely. If there is an in-line filter, either replace it or temporarily splice the line with a piece of metal tubing and clamps.

Check the fuel pump. If your engine has a fuel pump, the fuel may be blocked there. With the line to the carburetor disconnected, spin the engine with the starter and watch for fuel delivery. If no fuel appears at the pump outlet, a strainer or screen may be plugged at the pump. The pump diaphragm or valves may also be defective or stuck with gum. Only careful disassembly and inspection will reveal this (see Chapter Seven).

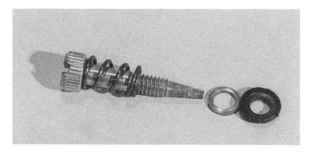

Always watch for the small washer and seal ring that are under the anti-vibration spring on most carburetor needle valves. Often they're down in a pocket and stay in place when the valve is removed.

Tank-Top Carburetors

This common type of fuel system is easy to spot because the carburetor body mounts on top of the tank. Such carburetors have extension tubes that go down near the bottom of the tank and pick up fuel through a screened pickup. Some of these suction-lift carburetors have a rubber/cloth diaphragm to aid the pumping action. The starting problem could be caused by the diaphragm and associated parts, and these are easily replaced. First, try removing the carburetor from the top of the tank and cleaning the screen and check valve if they're gummed up. You will need a new mounting gasket. You may have to take off the tank and carburetor as a unit and then separate the carburetor from the tank. This is not difficult and provides a chance to check the connection between the carburetor and the engine intake port. Not all carburetors have flanged mountings secured by nuts or screws. Some tank-top carburetors simply slide off a tube-like connection at the intake. This joint is sealed by a synthetic O-ring. If the ring has deteriorated, cracked, or been damaged, you've probably found your trouble. The only cure is a new ring.

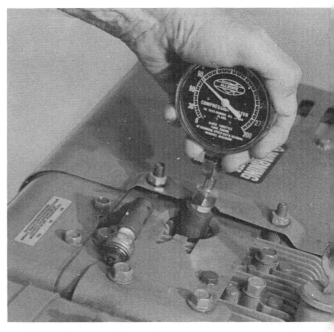

A few hearty pulls on this engine brought the needle up to about 90 psi—a fair indication that the valves and rings are in good shape.

To separate the carburetor body from the tank:

1. Remove the retaining screws and lift straight up.
2. If removal of the four obvious screws does not free the carburetor, do not pry or use force. Push the choke open and look down into the carb throat.
3. You'll see another screw down in the throat, and when you remove it the carburetor will lift off. For further details, see Chapter Seven.

Carburetor Cleanup

Let's assume that when you loosened the fuel line to the carburetor it was apparent that fuel was getting that far. The remaining trouble spots are commonly the main and idle jets. Some carburetors have no adjustment screws and cannot be cleaned without disassembly, but many have two knurled and slotted screws for the idle and main jets. These screws terminate in needle-like tips and extend into the jets. Screwing them in or out varies the amount of fuel that can pass the jets.

If you shoot carb cleaner directly into the jet, and the engine doesn't start and run immediately, you'll probably do well to remove the carburetor for a thorough cleaning as soon as possible.

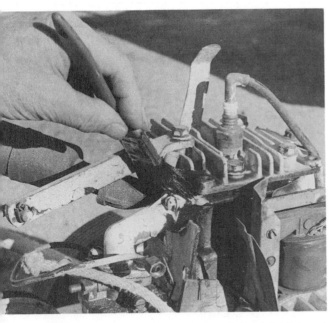

Painting some lube oil around the gasket joint (above) and pulling the starter showed enough fizzing on this engine to prove the gasket needed replacement. Also, if some head bolts feel tight and others feel loose (below), you'll need a new gasket and a torque wrench.

Removing the idle- and main-jet screws. Adjusting such screws is a fairly critical process, and since your engine had been running previously the adjustment positions are probably about right. So before removing the jet screws, first turn them in very carefully and count the turns, half turns, and quarter turns it takes to bottom them. Never force these screws and be alert for the feeling of bottoming. Make a note of the turns you needed and then back the screws all the way out.

Many such screws are sealed with small O-rings or rubber and brass washers where they enter the carburetor body. Watch carefully for these small parts since they're easy to lose. Also note the anti-vibration springs. They may not be interchangeable between the screws. In fact, on some carburetors it is possible to accidentally switch the main jet and idle jet screws. Keep track of them by pushing them through a piece of paper and labeling it.

Cleaning the jets. After removing the screws, examine them carefully for signs of stickiness or gum and discoloration at the tips. If you find it, you've probably located your trouble. Use carburetor cleaner or lacquer thinner and gently wipe the needle tips clean. This action may get you going, and once running the fuel may eventually clean out the jets down inside the carburetor. But if the jets are firmly clogged with gum, you will still have trouble.

One possible solution is to squirt a drop of carburetor cleaner into the jets with the needles out. Try using something such as a plastic cocktail toothpick to go down into the jet and carefully stir the cleaner around. Do not use wire or a drill. Now, replace the screws, adjust them to their original settings, and try to start the engine. There's an excellent chance it will run, though you may have to readjust the screw settings because the jets are not perfectly clean.

The main hazard with that procedure is that you are introducing a very strong solvent into the carburetor. If the engine starts and runs there's no problem because the fuel soon washes away the solvent residue. If the engine does not start, the solvent may find its way through the carburetor and seriously damage plastic or synthetic rubber over a few days time. The larger, float-bowl types are not likely to suffer greatly from solvent since their parts are usually metal, but the small diaphragm carburetors on chainsaws, trimmers, and the like may be vulnerable.

If you cannot get your engine to start and run after cleaning the jets from the outside, there is little choice but to disassemble the carburetor and find the source of the trouble. Here, careful work and cleanliness will probably get you by with flying colors; but unless you feel fairly confident you may be better off removing the carburetor and taking it to a repair shop. Again, you'll find more detailed guidance in Chapter Seven.

TROUBLESHOOTING COMPRESSION PROBLEMS

With the spark-plug wire disconnected, turn the crankshaft and listen carefully for hissing sounds at the carburetor inlet and exhaust outlet. Leaking valves are often signaled by hissing or blowing noises. If you hear no such noises but compression still seems low, try squirting a little engine oil into the spark-plug hole, replace the plug, then carefully turn the engine. If you feel the compression improve markedly, you've probably got a piston-ring or cylinder problem. If the engine was running reasonably well last week it's unlikely that it suddenly got sick sitting idle for a few days. But if you haven't run it for months, the rings might be stuck.

Before becoming too alarmed, try removing the blower housing and checking the head gasket. More fruitless hours are spent trying to start engines with blown head gaskets than might be imagined. That's because the owner very often does get the engine started, finally, and it seems to run fairly well. The reason is that during starting the engine is turning slowly and there's time for the compression to leak away. Once running the action is so fast that there's little time for leakage.

Checking the head gasket. One way to check a head gasket is to squirt or brush a little oil around the joint between the top of the cylinder block and head. The reason for removing the blower housing was to gain access to the area around the exhaust valve, the most likely point for the gasket to burn out. The exhaust valve will be located very near the exhaust-pipe connection. Crank the engine over vigorously and watch for bubbles in the oil. Any at all are a sure sign of gasket leakage.

Another quick check is to fit a properly sized box-end or socket wrench on each head bolt in turn. Pull firmly, but not so hard that you actually turn those bolts farthest from the exhaust valve—a less likely trouble area. If you test each bolt this way and come onto one or two that seem to yield slightly compared to the tightness of the others, you've probably found a bad head gasket. Although you may get the engine going just by snugging down the loose bolts, this is bad practice and can distort the cylinder wall and head and cause much more serious problems. The head gasket must be replaced, as shown in Chapter 13.

Leaking valves. Valves that are not too seriously burned or worn may allow you to start if you try long enough. Such valves won't get any better, and if the hard starting is followed a little later by a cracking or popping sound through the exhaust you'll know you have valve problems. In some cases a valve may simply be stuck open, but this usually prevents starting and shows up as no compression.

6

Manual Starters

Unless started electrically, most small engines today will have some form of recoil starter. These starters are handy, but they are also one of the first parts of the engine to give trouble. Often such trouble is caused by careless or over-emotional use, but ordinary wear and tear will also take its toll. Modern safety mowers, with blade brake/clutches may require a restart 10 or 20 times in the course of mowing the lawn. This compares to once or twice in the past and puts a greatly increased demand on the starter parts.

Recoil starters vary tremendously in design. On some you pull straight up, as is common on rotary mowers for safety reasons. Or, the action may be straight out in the plane of the flywheel. These are called *vertical-pull* and *horizontal-pull* starters respectively, and they differ markedly in their construction.

In this ball-type clutch, the badly worn notches in the engagement areas and gum-coated balls leave little doubt as to why this clutch didn't work very well.

GENERAL SERVICING TECHNIQUES

Be sure to examine the working parts carefully before disassembly. Try to analyze just how the parts are intended to function. Manually probe for sticking pivots or spiral engagements. Often you'll find all that's wrong is a little oil gum binding a pivot or sliding member.

If you decide to disassemble a small clutch as used in a chainsaw or yard tool for example, work over a shallow tray or cloth. Many such assemblies contain small springs and retainer clips that are easily lost. If necessary, mark the position of these parts carefully before total disassembly. Note, for instance, which way a dished, cupped, or wave washer goes. If it appears you'll need special tools, it may be better to seek dealer service. This applies especially to foreign-made engines.

RECOIL STARTER CLUTCHES

The old rope-wrap starter had one advantage in that it pulled free of the pulley notch and declutched itself. With recoil starters, since the rope rewinds under spring action, there must be some form of clutch or engaging device to couple the starter to the flywheel or crankshaft and then release instantly when the engine starts.

This means that a recoil starter has two basic mechanisms inside. One is the clutch and the other is the rewind mechanism for the rope. You can expect, eventually, to replace a rope or rewind spring or repair a clutch.

Ball-Type Clutches

The ball-type clutch has a cast housing with a series of notched pockets inside. The clutch housing

The wind-on rope starter on the old engine above had the advantages of simplicity and low parts cost. The modern vertical-pull recoil starter shown below is more convenient—and more complex.

threads onto the crankshaft and retains the flywheel. A second part, which runs inside the housing, has curved ramps. Between these two parts are steel balls. When the inner member is turned by the recoil pulley, the balls wedge between the pockets and ramps, locking the two parts together. When the engine starts, the balls move back into the pockets by centrifugal action and the reverse movement of the ramps.

In due course, the balls tend to become sticky and the contact faces of the inner member may become worn and notched or battered from the action of the balls. Repair requires a new clutch unit.

Removing a ball clutch. Your first job is to check the instruction manual for your particular engine or contact the nearest service representative to determine if the clutch is right- or left-hand threaded. You may also be able to see enough of the crankshaft end to determine this yourself. Your second job is to hold the flywheel so it doesn't turn while you exert the force needed to loosen the retainer.

Repair shops use a large C-shaped wrench that reaches around the wheel, in conjunction with a large strap wrench that grips the wheel rim. Unfortunately, these are repair-shop tools that may not be available when you need them.

An easy way to secure the flywheel is to remove the spark plug, lower the piston by turning the wheel, then thread some firm sashcord or clothesline in through the spark-plug hole. You can also try a device, called a Piston Restraining Tool, with a small Nylon cube molded onto each end of a tough Nylon strap. When you move the piston back up, the material will be compressed between the piston and cylinder head and will safely prevent the flywheel from turning.

With the air shroud and its built-in recoil mechanism removed, you'll see a square shank and a flat clutch housing with four small ears protruding from it. Lacking a special wrench that fits over these ears, you can easily make one from a short piece of pipe or tubing that fits over the clutch body (about 2½ inches diameter). Simply file, saw, or drill out four notches to engage the ears, then turn your home-made wrench with a pipe wrench, or drill holes to shove a bar through and turn it with that. For a fancier tool, weld or braze a top on the pipe, followed by a husky nut brazed or welded onto that. Now you can use a flex handle and socket to turn the tool as well as a torque wrench on reassembly.

Troubleshooting. It's possible with a ball-type clutch to pull the rope freely without engaging the crankshaft. Often a smart rap with the heel of your hand on the starter cover will jiggle a ball or two into engagement and you can start the engine. Again, you need a new clutch, but for now it may also be possible to pry the cover off the clutch and wash it clean, together with the balls, and have it function for some time. This does not require removing the clutch from the crankshaft; merely remove the blower housing for access. Do not oil the balls.

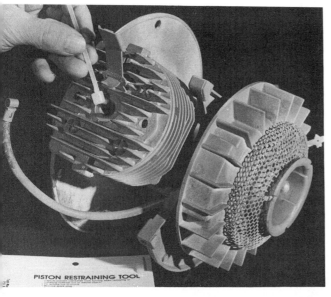

This small plastic tool is inserted into the spark-plug hole and prevents the piston in smaller engines from completing its upward stroke.

Tool above was homemade from a short piece of pipe. I welded a top and hex nut onto it. Four notches engage the ears of a ball clutch, as shown below.

Exploded view of a pawl-type rewind starter. Examine the pawls for wear, gum, and dirt; also check the center-cup contact surface, as well as the brake and rewind springs for cracks.

- Housing
- Handle assembly
- Rope
- Spring and keeper assembly
- Pulley
- Pawl spring
- Pawl
- Brake spring
- Retainer
- Starter pawl
- Retainer screw
- Centering pin

This dog-type recoil clutch has two dogs that move into engagement for starting.

Pawl-Type Clutches

Another popular clutch type uses one or two and sometimes three pawls—finger-like extensions that emerge from a shallow cannister-like housing and engage one or more indented steps in the surrounding drive cup. These parts can wear and lose their ability to engage solidly. Inspection is relatively easy, but be sure to wear eye protection and use care when disassembling these units.

Once the center screw holding the pawl housing has been removed, the pulley, rope, and rewind spring can be lifted out as a unit. Turn the pulley backwards gently until you hear or feel the bent-over end of the spring release tension. You can now lift the starter out safely since the spring is contained in a housing. If the unit does not lift out freely, the spring may still be engaged and further pulling may cause it to jump out of the housing.

Troubleshooting. Pawl-type clutch problems can usually be diagnosed by inspection. If the pawl or pawls emerge from the housing when the rope is pulled, carefully examine the engaging surfaces of the pawl and the cup in which it runs. Also, check the small spring under the pawl cover and be certain

that the cover retaining screw is tight. Be sure the pawl assembly and bushing are free-running and not gummed.

How a pawl-type starter assembly is attached to the blower housing will vary. On some engines the starter is retained by screws, on others by nuts, and on still others by blind rivets. In the latter case, removing the blower housing for starter work is easier than drilling out the rivets and replacing them. If your engine has a removable starter with three pawls it is important to center the starter when you replace it. To do this:

1. Install but do not tighten the screws or nuts.
2. Pull the rope enough to engage the pawls evenly.
3. Hold tension on the rope and tighten the fasteners.

Dog-Type Clutches

As in the old-time automobile crank, a toothed member on the crankshaft is engaged by another member, or "dog," with matching teeth. The teeth lock and engage firmly in one direction but ramp each other apart when the crankshaft starts to spin.

This dog-type clutch has a notched engaging member that picks up a tang beneath for cranking. Centrifugal force disengages the device when the engine starts.

Removing A Pawl-Type Clutch

1. Inside the blower housing, you'll see the tangs in the center hub that engage raised areas in the drive cup—shown at top removed from flywheel.

2. With clutch cover removed, the pawl is seen in its anchor in the hub. There is a small spring behind the pawl and another in the center of the hub.

3. Pull up carefully to disengage the pawl-starter recoil spring and pulley and lift them free. Hook end of the spring is visible in the center.

4. Slip the cover plate around to remove it for access to the recoil spring. New springs come coiled in a container and can be pressed right in.

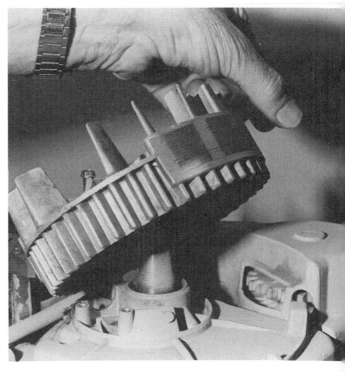

In this chainsaw starter-clutch, the two forked pivot arms at lower right engage the two small pins that are pointed out above.

Usually, coarse spiral threads engage the starter dog by moving it inward when you pull the rope. A variation on this is a design with finger-like engaging members rather than a matching dog. In most of these types centrifugal force throws the engaging fingers back away from the starter part once the engine fires.

In the above starter, the flywheel/ring gear and the starter gear—at right—are on the same plane. The starter gear shown below is at a right angle to the ring gear; when the rope is pulled it rises up and engages the ring gear.

Gear-Drive Clutches

Gear-tooth drive clutches resemble the toothed starter pinion and flywheel ring gear used with electric starters, although the parts are often made of tough plastic. They are most often used on vertical-pull starters where the force of your pull must be turned 90 degrees to rotate the flywheel.

In some designs the engaging gear or pinion is lifted by a spiral gear as you pull the rope. The teeth mesh with teeth cast into the flywheel rim, and both gears turn in the plane of the flywheel. A second type also has mating teeth in the flywheel rim, but the starter gear's axis pivot is set up so the starter gear turns at 90 degrees to the flywheel. When you pull the rope you also lift the gear out of a pocket and up against the flywheel.

Replacing Starter Ropes

The best time to replace a recoil starter rope is *before* it breaks. It's easier to replace the rope with the recoil spring already wound than having to gather the tools and setup needed to rewind it. Moreover, sometimes rope breakage causes the rewind spring to disengage from its seat and that means more work.

On Briggs & Stratton and Tecumseh engines, access to the rope is easily gained by removing the three or four screws that secure the blower housing and then removing the housing. If the rope is not broken, the spring will be coiled under some tension and the rope pull-handle will be drawn against the housing. To remove the old rope:

1. Secure the housing with two blocks of wood, a pair of C-clamps, and a vise. You could also try holding the housing between your knees or have a helper hold it for you.

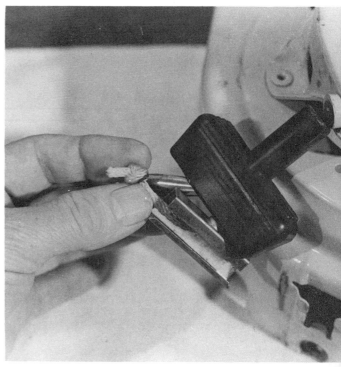

To replace a starter rope, begin by removing the rope from the pull handle. Pry out the end plate and you may find either a simple knot or fasteners.

2. Pull the rope out to the limit of its travel. This should align the knot location close to the outlet in the housing through which you'll extract the old rope and install the new one.
3. With the pulley in this position, use locking pliers or a C-clamp to lock the pulley against the housing so it cannot rewind under spring pressure. Excessive pressure isn't needed.

From here on the trick is to get the new rope through the passageway in the pulley and housing that is now occupied by the old rope. This may not be as easy as it seems since there can be a rope guide or hidden internal blockages.

One strategy is to nip off the old knot after pulling in a little slack in the rope. Then heat the old and new rope ends with a kitchen match or small torch until they melt and the ends can be merged together

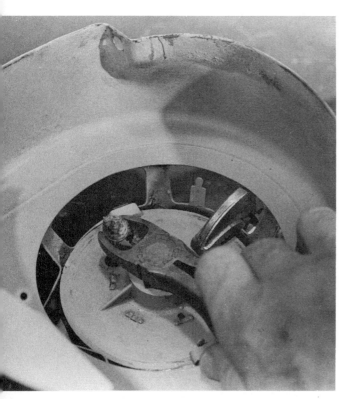

Pull the old starter rope out fully and clamp the pulley so the spring doesn't unwind.

at the point where you cut off the knot. While the nylon material is still soft, mold the merged area so it is the same diameter as the rope. Let it cool and then test the joint. If you did it right, the old and new rope will hang together solidly enough to let you carefully withdraw the old rope from the housing opening and pull the new one through behind it. Now all you have to do is measure off the needed length against the old rope—leaving enough for the knot in the pull handle, if used—install the pull handle, and knot the other end of the new rope in the pulley. When you ease off on the clamp, hold a hand on the pulley face and allow the new rope to be drawn slowly onto the pulley by recoil action.

Replacing a broken rope. If you were unfortunate enough to have the rope break, the method just described won't work because the recoil spring will have already dragged the broken end inside the blower housing. To get the new rope end through the housing opening and guide, make up a poking tool. Use about 6 inches of 1/16-inch steel wire, shape it to a point, and add a wooden handle. After heating the end of the new rope and shaping it into a sharp

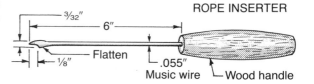

ROPE INSERTER

The tool shown above is easy to make and very handy for snaking a starter rope through the passage to the inner pulley. Below, the tool is used to poke the rope through the exit hole in the blower housing. This one seems to go easier working from the inside out.

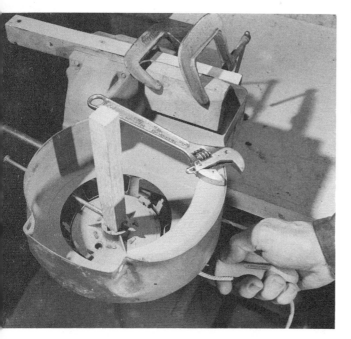

This homemade holding and winding setup maintains spring tension, locates the pulley for rope entrance, and holds the blower housing so your hands are free.

point, the tool can be forced into the rope just behind the tip and with a little maneuvering you can poke the rope end through the housing opening and past the guide inside.

All of the above assumes that you have already removed the broken stub of the old rope. Sometimes you can remove the stub by simply tugging on the knot and rotating the pulley. Other times, the rope is frayed and manages to jam in the pulley grooves so it can't be extracted easily. This means that you must gain access to the rope in the pulley.

On some engines, removing the screw in the center of the starter clutch and rewind unit will let you lift the unit out and remove the rope. On other engines, gently bend up one of the retaining tabs on the air shroud and carefully lift the pulley upwards just enough for access. Try not to lift the pulley any more than needed or you may disengage the recoil spring and have to start from scratch by reinstalling it.

Rewinding Recoil Springs

Once you've removed the old rope stub you must rewind the recoil spring and hold the pulley under tension while you install the new rope. In engines where the starter unit is held to the blower housing by screws or nuts, you can remove the unit and clamp it in a vise, but it's easier to clamp the entire housing with blocks and clamps.

Some instructions give you a specific number of turns for winding the spring, while others simply tell you to wind it until it is tight. If you use the correct length of new rope and the pull handle still dangles a foot from the housing when you're done, you didn't wind it tightly enough.

On engines where a square shank with beveled corners extends into the recoil clutch, the necessary winding tool is a piece of hardwood or metal about ¾ inch square that fits nicely into the recoil socket and extends comfortably above the edges of the blower housing. To make matters easier, drill the turning tool for an easy fit on a spike or piece of light rod. Now, with the housing firmly secured:

1. Start turning the tool with an adjustable wrench and slowly tighten until you feel the spring become firm. The pulley groove for the rope will probably be slightly past good alignment with the housing opening through which the rope must be passed.
2. Back off slightly, using the spike through the tool to prevent the spring from unwinding. Probe the passageway with a piece of welding rod or the like and adjust the pulley position as needed.
3. With a rope-poker tool, push the end of the rope in through the housing opening and into the pulley groove to the point where it can be pulled up into the pulley with a hooked wire or needlenose pliers.
4. Tug the end of the rope through the hole and, where applicable, tie a knot exactly like the old one. Attach the handle, carefully remove the tool and spike, and—applying some pressure on the pulley—allow the spring to slowly wind the new rope onto the pulley.

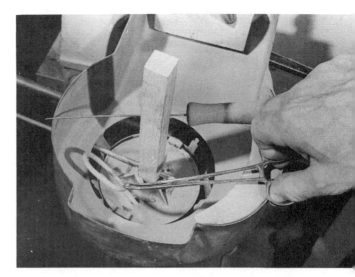

When the rope appears opposite the hole in the pulley, use a suitable tool to pull it up through the hole (above). Then tie a knot in the rope just like the old one and pack it down firmly, as shown below.

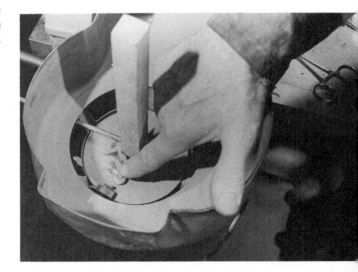

On other engines, with both the clutch and spring retained by a center screw, there is no way to use the wooden winding tool just described. The solution is a pair of locking pliers carefully adjusted for just enough drag to allow the pulley to be turned, yet keep the spring from unwinding. It is then fairly easy to put the tip of a screwdriver in the recess for the rope knot and use it to crank the pulley until the spring is tight. You'll probably find it easier with such starters to poke the rope through from the inside to the outside.

Rewind-Spring Replacement

When working with these springs, keep in mind that there is always the chance of a broken, jagged end escaping with the speed of a striking snake. Wear some form of eye protection and *be careful.*

The type of starter clutch/rewind unit with pawls has the spring—as supplied by Tecumseh—tightly coiled in a container and pregreased. Once the broken spring has been carefully removed, press the new spring out of the factory shipping enclosure and directly into the spring housing of the starter. You can also buy the spring and housing as a unit and simply replace the old spring assembly with a new one by slightly turning it so the lock tabs slide into place.

Briggs & Stratton engines use a spring that anchors at the outer end in a cutout in the blower/starter housing and at the inner end in a like seat in the pulley. To remove the spring:

1. Carefully pull out the end that locks in the housing and twist it to clear the cutout.
2. Bend up a tab on the housing to allow the pulley to be removed and freed from the other end of the spring.
3. Clean away any dirt and grit from the parts housing, straighten the spring gently by hand and oil it, then engage the spring in the pulley and replace the unit.

Use the spring-tightening procedure involving the turning tool and spike previously described. Take care to engage and seat the spring end in the housing cutout as tightening progresses. Once pulley and housing holes line up, install the rope as before.

Electric/Manual Start Kits

Many people cannot pull a manual starter because of physical limitations, particularly on small mower or tiller engines that may require considerable pulling when taken out of storage. Once started, the engine is usually easy to restart even if your pulling ability is limited.

An excellent solution is an adapter device that allows you to start the engine with a ⅜-inch electric

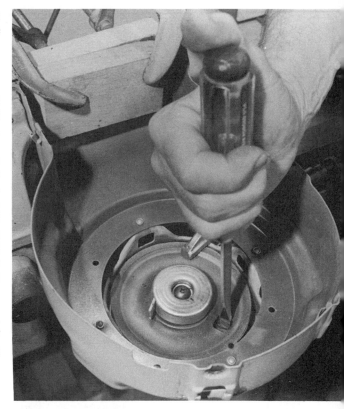

With this type of starter, wind the recoil spring by applying just enough pressure for clamping action with locking pliers and pulling the pulley around with a screwdriver.

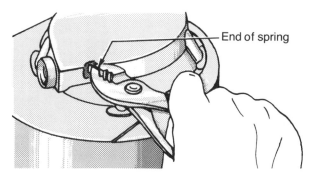

The outer end of this rewind spring anchors in a notch in the starter housing and must be drawn out and turned a quarter turn to be released. The inner end is similarly anchored.

drill, yet does not disable the conventional recoil starter. For a first start you simply plug in the drill and spin the engine rapidly and continuously. The kit for this device, called a Spin-Start, consists of a drive cup that attaches to the crankshaft with factory-supplied hardware. A cone-shaped rubber drive member has a shaft to fit the drill chuck. To use it, you push the cone into the drive cup and pull the drill trigger. Installation is a do-it-yourself job and seldom takes more than 30 minutes.

You can adapt your small engine for electric-drill starting with this easily installed device. Its rubber friction member attaches to your drill and engages a polished aluminum cup that mounts to the engine.

7

Servicing the Fuel System

Although you can disassemble a carburetor without really knowing how it works, it helps greatly if you do. Because if you understand the principles behind your small gas engine's carburetor and, for that matter, its other fuel-system components, you can usually put each back together again so that it functions like new.

WINTER AND SUMMER FUEL

To burn cleanly, gasoline must be thoroughly mixed with the proper amount of air—about 15 parts for every one part of gasoline. Ideally the gasoline would be in a vaporous, non-liquid form. But gasoline is not a single compound and is actually made up of 100 or more hydrocarbons. Some of these vaporize at relatively low temperature, while others may remain in a liquid state up to over 300°F. Gasoline blended for cold-weather starting has a higher proportion of hydrocarbons that vaporize at low temperatures than does gasoline intended for summer use.

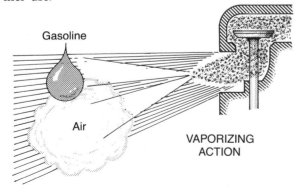

When warmed up and running normally, most small gas engines require a fuel/air ratio of about 15 parts of air to one of gasoline.

That can pose a problem for those of us with small engines. Although we may fill our automobile gas tank once a week or more, chances are we buy our supply of gasoline for our small engines and store it for weeks, months, and maybe up to a year if we add stabilizing agents. While left-over winter-blend gas will start and run your lawnmower in summer—possibly with some loss of power—don't count on starting your snowblower in the dead of winter with gas tailored for summer use.

Probably the biggest challenge is to the owner of a standby emergency generator. Such generators may get an exercise run only once a month, yet they require enough fuel on hand to provide several days of even intermittent running. Many owners have learned the importance of managing their fuel inventory to take advantage of the seasonal variation in fuel blend.

FUEL-DELIVERY SYSTEMS

Chapter Five lays out some general steps to making sure fuel is reaching the carburetor. Review these steps before doing any major carburetor work, since a problem in the fuel-delivery system will give many of the same symptoms as a faulty carburetor.

The simplest way to get fuel to the carburetor is a straight gravity system, with the fuel simply draining from a tank somewhat higher than the carburetor. The system is not satisfactory for many applications, however, because of varying operating angles. Also, in many riding mowers and tractors the fuel tank is remote from the engine and carburetor. And in some engine-generator installations, a separate, exchangeable tank may be used to permit remote refueling while the engine is running. Thus, a wide variety of fuel pumps have been developed to transfer fuel under these conditions.

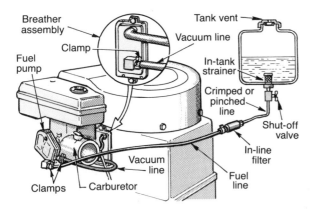

This fuel-delivery system uses a vacuum line at the breather to operate the fuel pump.

Mechanical Fuel Pumps

On some heavy-duty small engines, a small mechanical diaphragm pump similar to those used on cars is mounted on the side of the engine and driven from an internal cam on the camshaft or crankshaft. Some such pumps also have a small, lever-type manual primer for delivering fuel to the carburetor prior to manual starting. Primers can also save considerable battery energy with electric starting if the fuel has a long way to travel from tank to carburetor. When operating a manual primer you can usually feel the arrival of the fuel because the lever resistance firms up.

If a mechanically driven pump fails to eject gasoline when the engine is cranked, the pump may be dirty or have a plugged inlet screen, or it may need new valves and a pumping diaphragm. These items are available in kit form; to disassemble the pump, simply remove the screws around the edge of the diaphragm. But first remove the pump from the engine and inspect the lever shoe that rides on the internal cam. Severe wear means a new pump is needed.

Installing pump diaphragms. Two precautions are important when installing a mechanical-diaphragm pump. First, after hooking the pull rod that attaches the diaphragm to the actuating lever and before final tightening of the diaphragm screws, use the primer lever to pull a depression into the diaphragm. If the screws are tightened with the diaphragm flat, subsequent pulling action may tear the

diaphragm loose around the edges. If the pump has no primer lever, the actuating lever may be used. You can also install the pump with the screws loose enough so that you can watch the diaphragm as you turn the crankshaft until the diaphragm is depressed in the center. Then tighten the screws.

The second precaution involves properly locating the actuating lever on the cam. If when removing the pump a spring pressure is felt pushing the pump outward, the lever is on the high point of the cam.

1. Turn the crankshaft until spring pressure eases. This will make it easier to reinstall the pump, since you won't have to cope with spring pressure.
2. If the pump lever rides on a narrow cam on the crankshaft between the crankcheek and

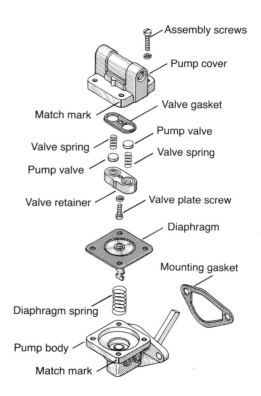

Typical diaphragm-type pump is actuated by a lever that rides on a narrow cam on the camshaft. The diaphragm is pulled down by the lever and returned by a spring, while two round valves work in opposite directions for fuel inlet and outlet.

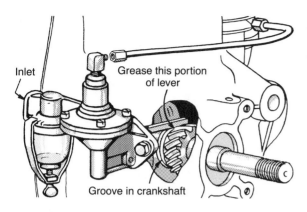

Installing a lever-type fuel pump requires careful placement of the lever on the camshaft or crankshaft.

the cam gear, you'll have to insert the pump straight in to prevent the lever from riding up on the gear.

3. Feel for proper positioning and slowly rotate the engine by hand to be sure the lever is correctly located on the cam.

Diaphragm/Pulse Pumps

A more common type of small-engine fuel pump uses the alternate positive and negative pressure pulses in the crankcase to power the pump. Because these pulses result from the piston movement, the engine must be running for the pump to deliver fuel. There is no manual primer to bring initial fuel from the tank, although the pulses from startup cranking are usually sufficient.

Diaphragm/pulse pumps are often mounted directly on the crankcase, with crankcase pulses transmitted through a small hole and filter to the backside of a synthetic-and-fabric diaphragm. A light spring supplies return pressure and the inlet and delivery valves are formed by small tabs or flaps in the diaphragm. Repair of this type of pump is self-evident from the pieces provided in the repair kit. The springs are rather small and delicate, and the entire pump should be removed and cleaned before disassembly. Be sure to check the screen and passage that connect to the crankcase. On disassembly, note the sequence in which the gaskets and diaphragm are sandwiched with each other. You could affect pump performance if the stacking order is wrong.

Other versions of the pulse/diaphragm pump may be located elsewhere, for example, on the carburetor. The crankcase pulses are delivered via a flexible line connecting to the crankcase at some point. In fact, some adapter kits provide a T-fitting into the dipstick tube as a point of connection.

CARBURETOR CHOKES

Referred to by the British as the "strangler," the choke performs exactly as its name suggests by partially choking off the air flow to the engine. Most chokes on float-bowl carburetors consist simply of

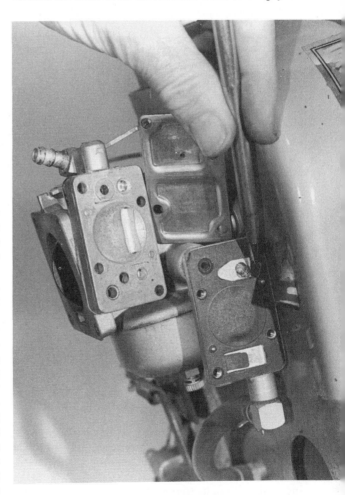

This simple fuel pump operates on crankcase pulses coming through the hose at bottom. Be careful not to lose the tiny spring shown at the tip of the pen.

With the air cleaner off, the choke valve is easily visible near the top of this chainsaw carburetor.

swinging plates or butterflies manually operated either by a lever on the carburetor or a control elsewhere on the machine. The choke is needed because at starting or cranking speeds the air velocity through the carburetor is not enough to induce the extra fuel required by a cold engine. This is because the gasoline tends to condense on the cold metal parts rather than vaporize. A richer mixture is needed until the engine warms up.

When partially or fully closed, the choke restricts the air flow through the carburetor and imposes more suction action as the piston moves downward on the intake stroke. This delivers raw fuel for starting, but as soon as the engine starts the engine will begin to labor, emit black smoke, and run roughly unless the choke is at least partially opened. In a minute or two even this mixture contains too much fuel, and the choke must be fully opened to let in more air. If your engine requires prolonged choking or choking while running and warm, the carburetor passages are somehow restricted, the float level is wrong, the carburetor is misadjusted, loose, or leaking at the mounting flange, or the fuel pump or fuel system is not delivering fuel properly.

Several types of carburetors boast automatic chokes, which operate in the following three ways.

Mechanical-Link Chokes

One widely used system employs a mechanical link between the throttle control and the choke. The choke valve itself may be either a sliding-tube or butterfly-plate type, but in each case setting the throttle in the *Start* position manually closes the

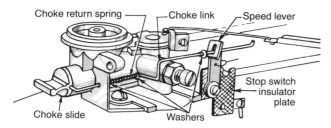

Mechanical-type automatic choke above is simply linked to the throttle-control lever on the handle of the machine. In the tank-top carburetor below, the force of the spring is balanced against intake-manifold vacuum for choking.

choke. To check it, simply remove the air cleaner and look down into the carburetor throat when the throttle control is moved. In most such designs it is important that the engine is installed and the throttle control connected. This is because the position of the Bowden wire control cable has a definite effect on the range of travel available to operate the choke. If the choke does not close properly it can usually be corrected by loosening the control cable clamp and moving the cable slightly to effectively gain the needed movement.

Vacuum Chokes

A second type of automatic choke—common on tank-top carburetors—uses intake manifold vacuum acting on a diaphragm inside the carburetor to open the choke when the engine starts and runs. Before starting, a light spring beneath the diaphragm pushes it up and a link between the top side of the diaphragm and a lever on the choke moves the choke butterfly to the closed position.

Thus, choke position is determined by the balance between the spring and manifold vacuum; if the engine slows down under load the spring will close the choke slightly and enrich the mixture. The parts involved are fairly delicate, especially the spring. A quick check of the choke action can be made by removing the air cleaner temporarily and, with the throttle closed, watching the choke valve while operating the starter or pulling the starter rope. The choke should respond to the intermittent manifold vacuum by opening and closing.

If the choke does not open and close several times as the engine is cranked over, a further check can be made as follows:

1. Remove the small screw securing the housing cover over the link and the choke arm.
2. Carefully try the choke arm for free movement and try to feel for spring action as the link and arm are moved.
3. If spring action cannot be felt, the spring—which is held to the bottom of the diaphragm by a clip—may have become disconnected or improperly oriented.

Vacuum-choke service. If there appears to be an internal problem, you'll have to remove the tank and carburetor assembly and lift the carburetor body off the top of the gas tank, which forms the lower part of the carburetor. Before removing tank and carb from the engine, however, carefully note the arrangement and placement of the small wire-control linkages in the governor and throttle hookup. These parts often have several holes in the arms, to which the linkages connect; make a sketch of where yours go now. Remember, too, to remove the hidden screw visible down in the carburetor throat when the choke is held open.

Once the carburetor is off, examine the diaphragm for signs of wear or damage. Then make a straight-edge check to be certain the two mating surfaces are flat. It is also possible, if the carburetor has been

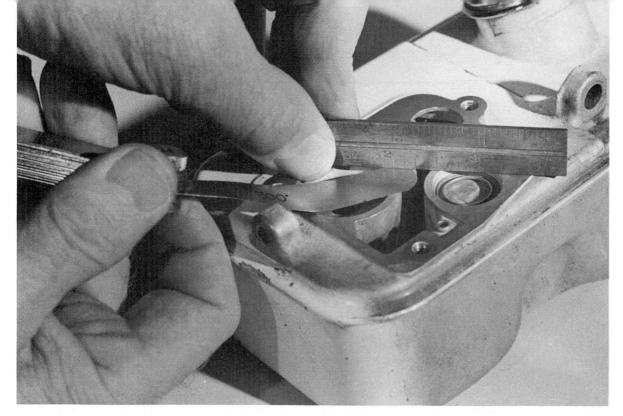

Here, an attempt to slip a .002-inch feeler under a straightedge is unsuccessful, proving the diaphragm-sealing surface atop this fuel tank is flat. Any warpage here will seriously affect engine performance.

previously disassembled, that the spring was damaged or bent out of position or not seated properly. In some cases the diaphragm may show signs of being wrinkled.

There are two important tricks to installing a new diaphragm in this type of carburetor:

1. Invert the carburetor, place the diaphragm in position with the spring vertical, and then carefully lower the tank section over the upper section, making sure the spring enters its well squarely. Hold the carb body and tank assembly together manually while you invert it to insert the screws.

2. Enter the screws just far enough to hold the two sections together. *Do not tighten.* Now engage the choke link and push the choke into the closed position, holding it there while you snug down the screws.

This preshapes the diaphragm with the spring extended.

Thermostatic assist. On some late-model engines used with mowers meeting safety compliance regulations, a vacuum choke may be assisted by a thermostatic element mounted on one end of the choke pivot shaft. A rubber line and connector feed hot gases from the engine crankcase in the normal breather function, but this heat is lead directly to a bimetallic coil on the choke. One end of the coil slips over an anchor pin and the other engages the choke shaft.

The choke chamber and coil may be inspected and cleaned by removing the rubber elbow. The shaft and spring assembly can then be withdrawn for cleaning or replacement. Since these parts are heat-sensitive, it is necessary to start and warm the engine thoroughly before making a final needle valve setting with the throttle stop adjusted for 1,750 rpm idle. *Be sure to observe safety precautions on mower engines.*

Full Thermostatic Chokes

Straight thermostatic chokes, actuated by a bimetallic sensing unit that responds to engine temperature, are relatively uncommon on small engines. On some of them, the choke lever swing is limited by a

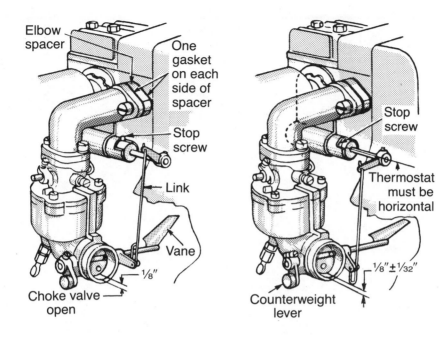

Thermostatically assisted choke uses a heat-sensing thermostat, in addition to a weight and air vane, to position choke valve for optimal operation. All linkages must pivot freely without any drag.

sliding linkage that moves up and down according to engine temperature. The choke is actuated by an air vane and a counterweight.

In addition to checking the choke plate for freedom of operation, be sure the choke is closed at normal room temperature and that the rotating thermostatic element is centered in its travel. Operate the engine until it warms up and observe the movement of the thermal element arm and the choke valve. When the choke is open, the counterweight should be horizontal and projecting towards the air cleaner.

Primers

Some engines use primers rather than chokes. Generally these are lawn-mower engines and are not expected to operate in cold weather. The primer is usually a soft, synthetic rubber button or sack.

Fully thermostatic choke uses a bimetallic coil spring to sense engine heat conducted to it via a hose connection from the breather.

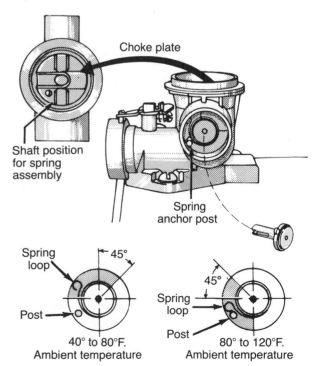

Primers deliver a shot of starting fuel into the carburetor as shown when you press them. Usually a single push is all it takes to start a cold engine and keep it running. Colder weather may require several pushes.

This float-bowl carburetor has a main-jet mixture adjustment screw at the bottom of the bowl, an idle mixture adjustment above, and an idle-speed adjustment screw on top near the throttle arm.

When pushed the bulb-like sack forces gasoline into the air passage for starting. This may be done in two ways.

One method delivers air from the bulb to a reservoir of fuel in the carburetor, and the air pressure forces the fuel out. The other system has a small vent hole in the bulb, which allows it to fill with air and is closed when you place your finger over it. A diaphragm underneath the bulb forces the fuel into the inlet.

In most cases a single push on the primer will permit starting and continued running, but in colder weather it may be necessary to tap the primer intermittently until the engine warms up.

CARBURETOR TYPES

The simple carburetors used on small gas engines all work on the same principles; only the appearance and parts are different. First, they must receive and temporarily store a small amount of fuel. Second, they must meter that fuel and transfer it into the stream of air to vaporize it as it enters the engine.

The storage function may take place in a bowl-like container, a depression in the carburetor body casting, or a tube extending down into the gas tank. The amount of stored fuel must also be regulated constantly as some of it passes into the engine and more is added from the fuel tank. This may be done by a float and needle valve in float-bowl carburetors, and by a diaphragm and needle valve or by a series of flap-like valves in diaphragm carburetors. The float-bowl type is most easily understood and, for those unacquainted with carburetors, it is probably the best place to start.

Float-Bowl Carburetors

The float chamber may be a metal casting, a stamped metal bowl, or even a plastic bowl, but it is always prominent enough to distinguish this type of carburetor. The fuel inlet fitting to which the fuel line attaches leads directly into the top of the float bowl. Sometimes you may find a fine-mesh screen at this point. Directly below the fuel passage you'll find a screwed- or pressed-in valve seat. This seat may be entirely of brass or it may have a tough, plastic insert.

Opening and closing of the fuel passage in the seat

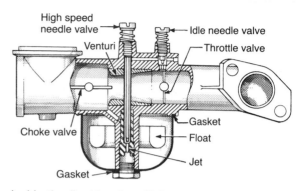

Inside the float bowl you'll find a float, in this case a hollow metal ring (photo below, bottom-right). Also seen below is the float needle valve and pivot pin, and up in the cavity at center you can just see the float needle seat.

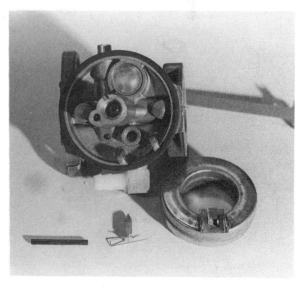

is controlled by a needle valve, commonly referred to as the float needle. It, too, may be solid brass or stainless steel, or it may have a resilient synthetic tip. The float needle is more or less freely mounted on a pivoted arm attached to a float. The float may be plastic or cork, or a soldered-and-sealed hollow brass ring, but it must resist gasoline and moisture.

When sufficient gasoline has entered the float chamber, the float will rise until the float needle seats in the valve seat and stops the flow. As the engine uses fuel, the float will drop slightly and allow more fuel to enter. Occasionally engine vibration or jostling such as on a mower or tractor will

How A Float-Bowl Carburetor Works

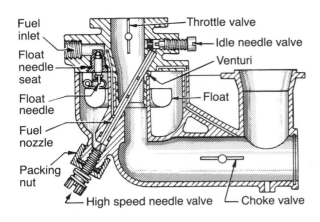

Fuel inlet — Throttle valve — Idle needle valve — Float needle seat — Venturi — Float needle — Float — Fuel nozzle — Packing nut — High speed needle valve — Choke valve

1. As the float rises, the needle closes off the inlet fuel-needle seat. As the level in the bowl drops, the float allows the needle to open the seat passage.

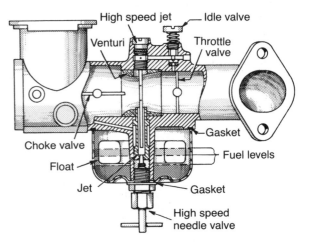

High speed jet — Idle valve — Venturi — Throttle valve — Choke valve — Gasket — Float — Fuel levels — Jet — Gasket — High speed needle valve

2. Although internal parts and passages differ greatly, fuel level relative to jets and other metering parts is critical. Note location of high-speed fuel pickup.

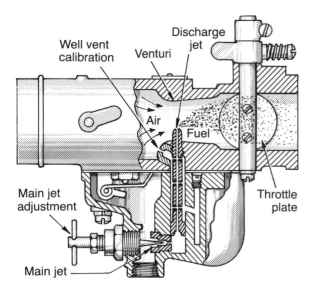

Well vent calibration — Venturi — Discharge jet — Air — Fuel — Throttle plate — Main jet adjustment — Main jet

3. The *venturi*, or throat, causes intake air to increase speed and reduces pressure. The result is that the fuel charge flows into the air stream.

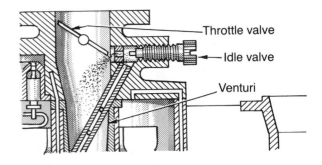

Throttle valve — Idle valve — Venturi

4. Idle jet comes into action when the throttle plate is almost closed. Idle speed is controlled by the small screw on the outside of the carburetor.

temporarily upset perfect inlet control, but the action is accurate enough for practical purposes.

Float-valve wear. An engine will usually start and run even if the float needle and seat are somewhat worn or misadjusted. On gravity-feed systems, however, if the needle and seat don't seal perfectly, gasoline is free to drain from the tank and drip out of the carburetor. If you store such an engine in a garage or shed, the fumes that accumulate could spell disaster.

Another aspect of a worn or improperly adjusted float needle and seat relates to how your engine performs. When the carburetor is designed, the metering and discharge jets as well as the internal passages are located so the fuel level in the bowl matches them. If the fuel level is too low, metering is impaired because the fuel must be lifted too far. If the fuel level is too high, excess fuel will be delivered.

Fuel metering. Fuel metering starts with the control of how much air enters the engine. This is normally done by the throttle plate, often called a *butterfly valve.* If you look down into the carburetor throat with the choke plate open you'll see a circular plate that can be pivoted from an almost on-end, wide-open position to an almost closed position. Throttle-plate position may be manually controlled, as in your car or a chainsaw, but is more commonly managed by the governor.

The venturi. Look directly behind the throttle plate and you'll see that the diameter of the air passage becomes smaller. This constricted area is called the venturi and its purpose is to force the air to increase its velocity at that location. The laws of physics and air flow are such that an increase in velocity is accompanied by a decrease in pressure. This means that the rapidly moving air going through the venturi is momentarily at a pressure lower than that of the atmosphere. Meanwhile, the gasoline in the carburetor, which is at atmospheric pressure, will try to find its way through the internal passages to any outlet that leads to the low-pressure venturi area. This establishes a flow of gasoline through the carburetor to a discharge into the rapidly moving air stream.

Main jets. To accurately tune the fuel flow for different operating conditions in float-bowl carburetors, restrictive jets are needed in the carburetor passages. These are usually threaded fittings with carefully sized holes, into which are screwed adjustable needle valves. One jet, usually called the main jet, may be located at the bottom of the float bowl.

If you remove an adjusting needle, use enough light to spot the small rubber seal ring and the washer behind it. Be sure not to lose these parts.

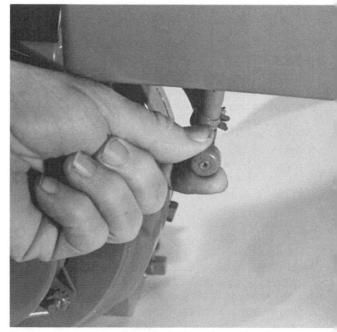

Before removing the float bowl, close the tank valve or pinch off the line with a clamp.

It affects high-speed and load performance by metering the fuel flow into a tube, also a form of jet, which leads to the venturi. Most carburetors have some form of tube leading to the point where fuel discharges into the low-pressure air stream. The tube may be plain, or it may have a number of very fine holes drilled through it. In most cases the holes serve to bleed air into the fuel to aid vaporization.

Idle jets. A float-bowl carburetor's idle system and passages are usually separate from the main delivery system. The idle fuel outlet in the carburetor throat may consist of one or two very small holes located so that an almost-closed throttle plate never quite cuts them off. Thus, in addition to the main jet, you'll also find an idle jet with an adjustable needle valve, which commonly enters the carburetor at the top or side and controls the idle fuel flow.

Unless your machine has fixed jets—as do some lawn mowers and other machines with more or less fixed speeds—you should find a knurled screw in the center of the nut that retains the float bowl, or on top of the carburetor directly over the fuel discharge tube. In both cases this is the main-jet needle valve. Screwing it in slightly will restrict the jet passage and reduce or *lean out* the fuel flow; backing it out allows more fuel to flow, resulting in a *richer* mixture.

The idle-jet needle valve has little or no effect on high-speed and load performance. But with the throttle closed you should be able to similarly adjust the mixture for smoothest running at idle. Idle speed is controlled by an adjustable stop screw that establishes the limit of throttle closing.

Carburetor seals. Always remember, when we talk about carburetors, that we are dealing with extremely minute amounts of gasoline on any given intake stroke. This means that leakage of either fuel or air will upset the metering and cause erratic operation.

Some leaky seals, such as a float-bowl gasket, may be evident simply because the gasoline leaks out. Others are more subtle in their effect and may cause unstable running speed, high idle speed, overspeeding and surging, or running on even though you close the main jet needle. Seals may be identified as small O-rings, or small, flat synthetic rubber washers. Look for them when you disassemble the carburetor, and replace them with new ones. If old seals must be reused, do not soak them in carburetor cleaning solvent.

Carburetor-flange fasteners shown here are by no means the most awkward you may encounter.

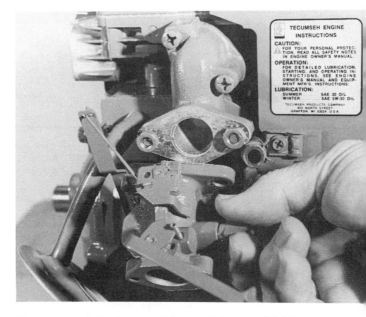

Carburetor and governor linkages take careful handling. Always note and record which holes the links were in originally.

In other carburetors sealing may depend on secure clamping of a diaphragm between two flat surfaces of the carburetor body. If the mating surfaces aren't perfectly flat, leakage between different sections of the carburetor may produce problems that are hard to track down.

Cleaning A Float-Bowl Carburetor

This is almost a routine job, although it can usually be avoided by running the carburetor dry before storage. It is not difficult if you follow an orderly procedure.

Float-bowl removal. Again, be sure to first note the location of all wire-control linkages; *do not try to bend them free now.*

1. Disconnect the fuel-line connection and allow the fuel to drain into a container if there is no fuel shutoff valve.

2. Remove the air-cleaner backplate and loosen the nuts or capscrews holding the carburetor flange to the engine. You may note that these fasteners have slots in them; that's because they are often located in a miserable spot to get at with a wrench. Once they're loose it's easier to diddle them out the rest of the way with a screwdriver. You'll probably have to back out one until it interferes with the carburetor body, then do the same with the opposite one, pull the carburetor back off the flange a ways, and proceed until both sides are free.

3. Once the carburetor is free you can tip and cant it to disengage the linkages without bending or distorting them. Do this gently.

4. With the carburetor free, wash it externally with a small stiff brush and lacquer thinner or carburetor cleaner. The idea is to get rid of all the tiny bits of grit and dirt on the outside before you actually disassemble it.

Carburetor and float-bowl disassembly. From here on the rule is to do all things gently. Do not smoke or work near a source of flame, since some raw gas is still present inside the carburetor. Find a good workbench with adequate lighting and cover the work surface with clean paper. The small metal trays used for TV dinners make fine parts and wash trays. A small watercolor brush or a source of compressed air such as the pressure cans used in darkrooms is also handy.

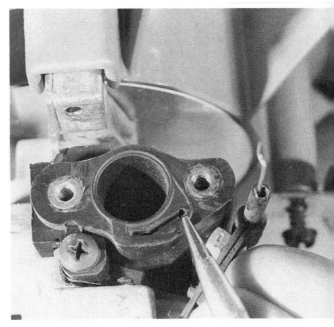

If you remove a carburetor and encounter a mounting flange like this, be especially cautious to note the exact stack-up of the gaskets and communicating holes.

1. Before removing the external mixture-adjustment screws, be sure to count and record how many turns and half turns it takes to screw them all the way in; this lets you return them to the same setting on reassembly. Once they're out, separate them for location. If the screws show any trace of gum, soak them in carburetor cleaner. *Do not soak the seal rings.*

2. Scrape away any residual flange-gasket material. Examine the flange and note any small drilled passages that match holes in the intake mount on the engine. Only a few engines have these but they must have a gasket with matching holes on assembly.

3. Hold the flange against a flat surface. If you have a fresh piece of No. 240 wet-or-dry abrasive paper, place it on the flat surface and draw the flange across it firmly. If the flange is bowed or there are high spots around the bolt holes, as shown by bright areas, dress the surface flat with the abrasive or a flat file. Wash away all cuttings.

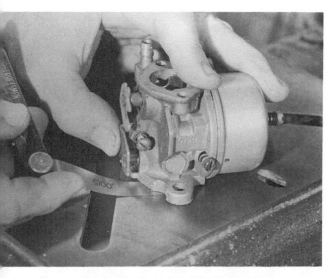

Above, a .0015-inch feeler will slip under the central part of the mounting face. Light rubbing on a piece of fine abrasive on a flat surface will usually correct this warpage problem.

4. Before going any further, test the freedom of the throttle-plate and choke-plate pivots. If they bind, squirt carburetor cleaner around the shafts and work them free. You may also find excessive wear to the point where the shafts are sloppy in the carburetor body. This probably wouldn't keep the engine from starting, but a badly worn choke pivot will allow fine dirt to enter the engine and a loose throttle shaft will cause erratic running and make mixture adjustment difficult.

5. Check the tightness of the small screws that hold the throttle and choke plates on the shafts. You're probably better off not removing these since failure to tighten them properly on reassembly can result in a screw going into the engine. More than one piston or valve has been damaged that way.

6. Remove the float bowl retaining screw. There's usually a fiber washer under it and this should be replaced with a new one. You may also find that this screw is actually a jet, with a number of small holes and perhaps a seat for the mixture adjustment needle. If so, place it in cleaner to soak.

7. Gently loosen the bowl-to-body gasket and try to carefully lift the bowl free. Usually it will come off without a problem, though some carburetors have a jet or metering tube that extends into them and could be bent if you get too rough. If you encounter resistance, search around for the screw that indicates such a part and remove it before continuing.

8. With the bowl free, examine the bottom of it and the surrounding parts. Is there silt, rust, gum, rubber hose scraps, or white or green corrosion? Such material must be flushed and soaked away gently until the parts—including the often-minute drilled passages—are bright and clean. Use a brush, solvent, and, if possible, compressed air to do the job.

I don't recommend removing the small, screwed-in brass jets that you find in some carburetors. Usually these are seated very tightly, and you'll almost certainly damage them with ordinary screwdrivers; carburetor repairmen use special jet drivers. Also, I wouldn't recommend punching out an expansion plug. These little discs, together with lead balls, are used to plug the access points where the factory drills passages in the carburetor body. If

Float-Bowl Carburetor Disassembly

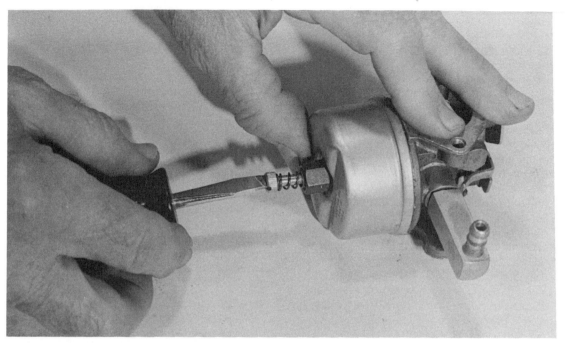

1. Begin by removing the carburetor needle valve at bottom. Carefully turn it in until it bottoms. Count full and fractional turns and write them down. Then back the needle valve out for cleaning or other service.

2. With a tray beneath to catch spilled gasoline, loosen nut and gently lower bowl. Float will swing free.

3. If it contains thick residue as shown, more extensive carburetor cleaning is needed.

such a plug leaks or seeps gasoline, a light tap with a blunt drift will seal it. Or, you can clean the area and put a little epoxy cement over it.

Float-needle and seat service. Because the float needle is subject to constant jiggling and vibration, both the needle and seat eventually wear. To service them:

1. Push out the float pivot pin and note the position of the light spring clip used in many carburetors to stabilize the needle.
2. Examine the float needle tip for signs of wear. If you see a ring of wear around the tip, the needle and seat must be replaced. Many float needles have tips of synthetic material and must not be soaked in strong cleaning solvent.
3. Check the inlet opening to the float-needle seat for accumulated dirt, gum, or even small particles of fuel hose or rust. In some cases you may find a small, fine-mesh screen in this opening. This can be drawn out and soaked in solvent.

4. Before reassembling the carburetor, the float height must be set to the specification in the manual for your particular engine. This may be done by measuring with a ruler from the top of the carburetor to the upper surface of the float, or by slipping a twist drill of the proper fractional size into the gap. Do this with the carburetor inverted so gravity brings the float arm into contact with the needle and the needle rests in the seat. Do not attempt to force the float. Make adjustments by minute bending of the little tab that contacts the needle. Use needlenose pliers.

Reassembly and adjustment. Reassemble the carburetor exactly the way it came apart using new gaskets but no gasket cement. Be certain that all the parts you soaked in cleaner are clean and dry.

1. Install the idle and main metering needles and their seals so they just bottom and then back them out to their original settings.

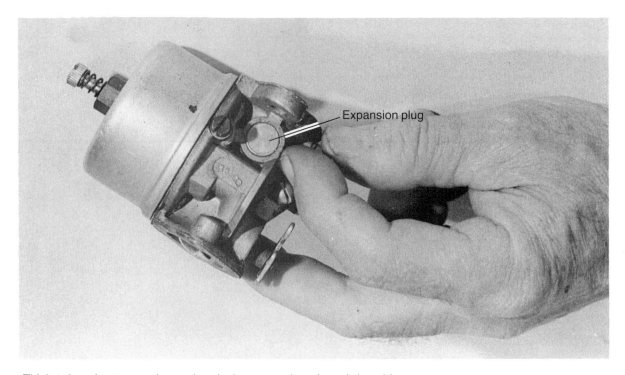

Think twice about removing and replacing expansion plugs. It is seldom necessary.

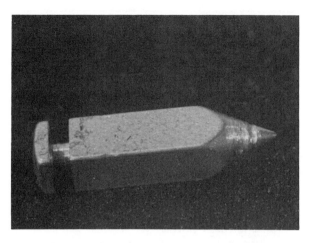

Not all float needles will show wear as bad as this, but even a little wear causes leakage and erratic operation.

Removing the carburetor almost always damages the gasket beyond re-use. Replace it on reassembly.

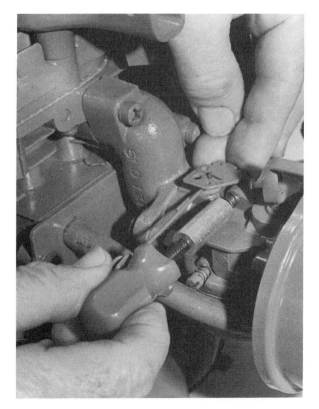

The idle-stop screw on this carburetor controls the low-idle speed. With the engine idling, hold the throttle arm against the stop screw and adjust for desired speed.

Specifying float adjustments is by drill size. With the float needle seated by the weight of the float, the proper drill shank should just slip in.

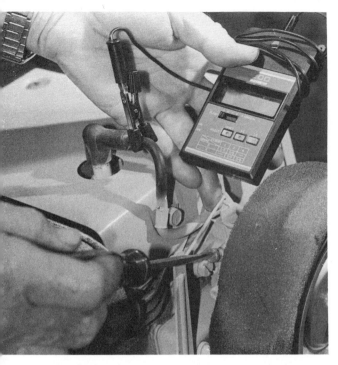

A digital tach will show the effect of even minute mixture adjustments, but for practical purposes your ear may be just as good.

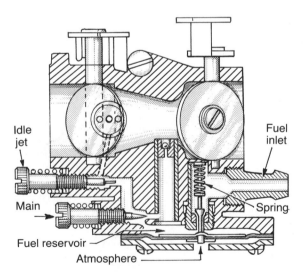

Typical diaphragm carburetor without a fuel pump essentially duplicates the action of a float carburetor. Fuel enters the inlet and is restrained by a valve that moves in response to a diaphragm.

You can't improve on this setting until the engine is running.

2. Mount the carburetor on the engine and engage the control links as you maneuver it into place. A thin film of gasket cement may be used on the flange gasket but it should not be needed.

3. Connect the throttle and choke controls and tighten their clamps. Make sure you have full travel of both choke and throttle plates. Your engine should now be ready to start.

If it is impossible to load the engine while adjusting the carburetor, as it is on a tractor, you'll have to adjust by trial and error until you get the best setting. Make only small adjustments at a time. If the engine is on a rotary mower or other machine where a rotating blade or other part constitutes a load, plan your moves and positions very thoughtfully to avoid injury.

1. Make a preliminary main-jet adjustment. If the engine was running before, and you recorded the turns the mixture needle was open and then returned it to the same adjustment, you should be close enough to start and warm up the engine. If you didn't do this, try starting with the main jet 1½ turns off the seat.

2. Let the engine warm up and then turn the needle clockwise (in) to lean out the mixture. Note when the engine starts to slow down and run unevenly.

3. Turn the needle counter-clockwise (out) to richen the mixture until the engine again starts to run unevenly.

4. Turn the needle clockwise again towards the lean position until the engine smooths out. This point will probably be about halfway between the rich and lean points.

5. Close the throttle control to idle and hold the throttle plate lever against the idle stop screw with your finger. If the engine is not idling smoothly, repeat the procedure used for the main jet setting and adjust for a smooth idle about halfway between the rich and lean points.

6. Check the idle speed with a tachometer if possible. Lacking a tach, you may have to judge by ear and previous experience. A commonly used idle setting for small four-

Above, button at center of the diaphragm (foreground) bears against the end of the needle, seen at the center of the reservoir chamber. Below, close-up of diaphragm carb shows the needle, spring, and needle seat that screws into carburetor body.

stroke engines is about 1,750 rpm. Do not try for an extremely low idle speed. Adjust the final speed by turning the stop screw against which the throttle plate lever seats at idle.

7. Quickly open the throttle from idle to full speed. The engine should accelerate quickly and without hesitation. If it seems to stagger or tends to die on quick acceleration, try opening the main jet needle a fraction of a turn at a time. It may also help to increase the low idle speed somewhat.

Diaphragm Carburetors

There are three basic types of diaphragm carburetors. One type, which has the carburetor body mounted on top of the tank, is somewhat different and will be discussed later. Another type commonly used on chainsaws is actually a combination fuel pump and carburetor.

For the moment, consider the simplest form of diaphragm carburetor. It performs all of the functions previously described for the float-bowl carburetor, but because it is not as vulnerable to off-level positions that would displace a float, it fits many small-engine applications better. As with the float carburetor, there must be a regulated reservoir of fuel that is constantly replenished as the engine runs. This reservoir is smaller than in most float-bowl carburetors, and is simply the space between a flat diaphragm and the bottom of the carburetor body.

As with the float carburetor, a needle valve and seat control the entry of the fuel by gravity into the storage area. The closing force, acting downward on the needle valve, is provided by a closely calibrated small, light spring. This equates with the upward actions of the float in preventing fuel flow.

Unlike the gravity action of the float, however, the diaphragm is forced upward to lift and open the needle valve by atmospheric pressure. One side of the diaphragm is vented to the air and the side containing the fuel is exposed to manifold vacuum via a small passage from the carburetor throat. Atmospheric pressure, being greater than the reduced pressure in the throat, overcomes the light spring pressure and pushes the diaphragm against the needle to allow fuel to enter.

Straight diaphragm carburetors are more sensitive to anything that might influence the air pressures involved. A leaky carburetor-mount gasket, leaking air-cleaner gaskets, or leaks around the throttle and choke shafts, for example, may upset the balance between the diaphragm and its spring and the pressures controlling the fuel flow. But the simplicity of repairing and replacing the few parts

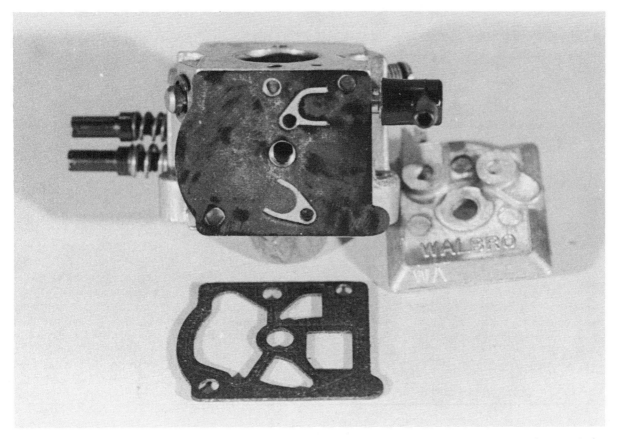

This still-smaller diaphragm carburetor is widely used on chainsaws. Shown above is the pumping side that brings fuel from the fuel tank into the carburetor, to be mixed with air and burned in the combustion chamber.

involved make this type of carburetor an attractive candidate for home repair. Diaphragm carburetors may have one or two mixture adjustment needles, which are set the same as with float carburetors. Again, the seals around the needles should be in good condition or replaced with new to avoid leaks.

Fuel-pump/diaphragm carburetors. The many positions into which chainsaws, hand-held tillers, and the like may be placed rule out gravity-feed fuel systems and put a premium on lightness and compactness. Carburetors for these engines are typically fed by a flexible pickup tube in the gas tank. These tubes are fitted with intake screens that also serve as weights so they follow the gasoline as the tank is turned and tilted. A built-in pump induces fuel flow from the tank to the carburetor.

You can recognize fuel pump/diaphragm carburetors because the side the fuel line enters has an unvented cover plate, while the opposite side cover plate has a small vent. The vented side admits atmospheric air pressure to the bottom of a diaphragm to regulate the fuel supply as described for straight diaphragm carburetors. Although there are many detail variations, these carburetors are built by only a few manufacturers, and even foreign-made engines are commonly equipped with them.

A number of points are important to any repair or troubleshooting of pump/diaphragm carburetors. First, the fuel pumping side of the carburetor will contain a diaphragm about the size of a postage stamp. This reflects the need for compactness and also the fact that the actual amount of fuel delivered to the engine is quite small and metering it is a precise operation.

The limited size of the pump means that several flexings of the diaphragm may be required to bring

For pumping action to occur, the crankcase pulses must find their way through a short labyrinth of holes and passages in the gaskets and spacers.

fuel from the tank to the carburetor if the fuel line is dry. That may also mean having to pull the starter rope more than the normal two or three times.

You'll probably find, on disassembly, that each gasket and spacer has a very small hole and perhaps a curved passage that looks much like a worm hole in an apple. All of these passages ultimately connect the engine crankcase with the pumping diaphragm chamber, and as the crankcase pressure fluctuates with the movement of the piston the pumping action is established. Be sure to keep the gaskets and spacers in the order and orientation in which you removed them so that the holes and channels align. Otherwise, there will be no pumping action and the engine will not run. Usually you're better off avoiding the problems of frayed gasket edges or bits of broken material plugging the holes by replacing the used gaskets with new ones on reassembly.

One other point should be checked. Since the small passage that operates the pump connects directly with the crankcase, it is possible for thick residues from two-stroke fuel to seep in and plug the hole. This most often happens when the engine is

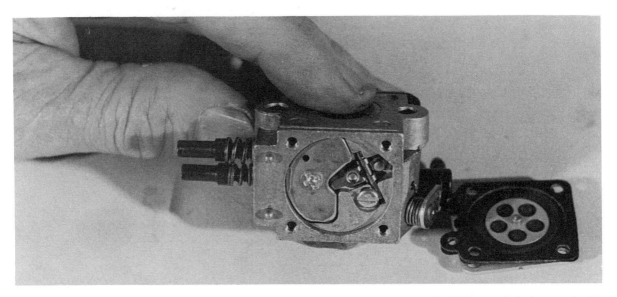

Metering side of the dual pump and carburetor unit. In this one the diaphragm acts against a pivoted arm that raises and lowers the fuel-inlet needle in its seat. In this way, the fuel supply is regulated.

stored in an unusual position such as on a chainsaw standing on its handle in your car trunk.

Replacing pumping diaphragms. The pumping diaphragm should be replaced whenever the carburetor is disassembled, but since its main parts are two small flap-like cutouts that act as valves, you may be able to reuse it. Hold it up to the light and look for any pin holes or thin spots. If you are going to clean the carburetor with carb cleaner, set the diaphragm aside since such solvents will destroy it. If you do find a badly deteriorated diaphragm, there's a chance your gasoline dealer has been selling you fuel laced with methyl alcohol. Many small engines have suffered from alcohol fuel blends in recent years.

On the carburetor side of these small units you will find essentially the same parts—diaphragm, needle, and spring—that are used in larger carburetors. These parts are still smaller and more delicate, however, and you may find a pivot arm or lever that works against the needle at one end and the spring at the other.

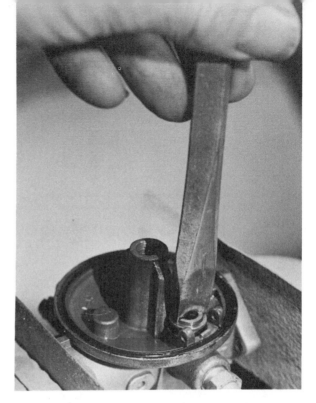

To remove and replace a float inlet valve seat, use a wide screwdriver carefully ground to fit the slot in the valve seat (above). Inspect the passage area between the fuel-line fitting and the seat for dirt and debris, then screw the seat into place and tighten firmly as shown in the photo below.

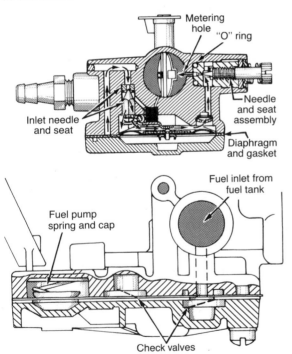

This single-diaphragm carburetor incorporates pumping action with the metering diaphragm. Pulses from a two-stroke crankcase cause the pumping section to deliver fuel via two check valves.

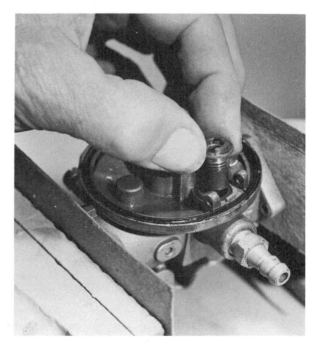

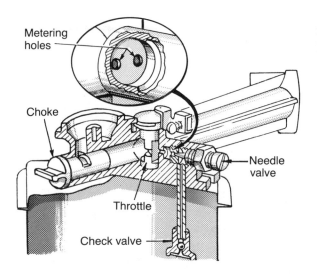

Metering holes

Choke

Needle valve

Throttle

Check valve

Single-tube, tank-top carburetor uses negative pressure of engine intake to elevate fuel from the tank into the pickup tube. A small, ball-type check valve prevents fuel from draining back.

Truly, working with these parts is almost like working with watch parts. A good headband magnifier is helpful, especially when looking for wear, grooving, or damage to the needle and seat or the mixture-control needles. A magnifier may also help you spot bits of dirt you might otherwise miss. It is also an excellent idea to work on these tiny carburetors over a shallow pan of some sort so that any dropped parts will be contained. A tiny spring popped out onto the shop floor can take hours to find—if you find it.

All of the metallic parts—as well as the carburetor body, any internal screens, the mixture needles, and the passages—should be flushed and blown out bright and clean. If the fuel-control needle has a non-metallic tip, do not expose it to the cleaner.

Disassembling single-diaphragm-type pump/carburetors. Although similar in principle to the double-diaphragm fuel pump/carburetors just discussed, the carburetor type found on some larger two-stroke engines uses a *single diaphragm*. You'll also find that it is of distinctly different appearance and construction.

You can recognize these carbs by the fact that they are flat on the bottom and have a single diaphragm sandwiched between body and cover plate. To inspect or repair this type of carburetor, it must

first be removed from the engine and inverted. Then proceed as follows:

1. Before removing the bottom plate, place the unit over a shallow pan or cloth to catch the small springs and other parts that may be released. You'll find a spring and retainer cup on the pumping side, plus two small valves that may be fitted with small springs.

2. Note that the pumping chamber and fuel-metering chamber are sealed off and divided from each other solely by the clamping action of both the bottom plate and body on the diaphragm. *This central seal area is critical to carburetor performance.* Check with a straightedge to be sure the clamping area is no more than .003 inch lower than the mounting face for the cover.

Replacing the inlet valve seat is somewhat more difficult than the screw-in type seat. First you must use a small hook to lift out a retainer ring and then another hook to lift out the seat. The new seat is pressed in until it bottoms and the retainer pressed back in above it. The Briggs & Stratton manual specifies pressing in the latter to 5/16 inch below the cavity surface. Your best bet when servicing this type of carburetor, however, is an instruction manual covering your engine make and model, since a number of varying techniques and details must be observed.

Among these are the removal and replacement of the check valve and the adjustment of the diaphragm hinge-lever height, much like a float height. These carburetors may be fitted with both a primer connection for a remote pushbutton primer as well as a butterfly-valve type choke. Also, the butterfly choke may be arranged for automatic operation when the throttle control is positioned for starting.

Tank-Top Carburetor Service

For many years, Briggs & Stratton engines have had their carburetors mounted directly on top of the fuel tank. There are many detail variations among the several models of tank-top carburetors, but these units are usually fairly simple in construction and can be repaired by the engine owner with kit parts.

One characteristic common to suction-lift tank-top carburetors is that they deliver the fuel from the tank through a brass or plastic tube that extends

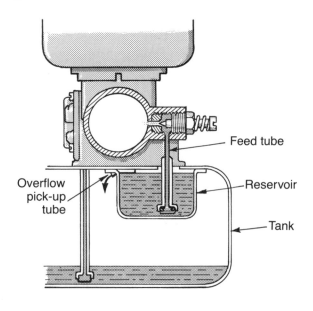

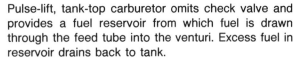

Pulse-lift, tank-top carburetor omits check valve and provides a fuel reservoir from which fuel is drawn through the feed tube into the venturi. Excess fuel in reservoir drains back to tank.

You can replace the pumping diaphragm on this tank-top carburetor without removing the carburetor. Turning the engine on its side helps keep diaphragm and spring in their place.

down near the bottom of the tank. In the simplest configuration, this tube will have a screen at the bottom and will contain a small ball-check valve. Negative pressure in the venturi passage causes air pressure in the tank to force the fuel up into the tube and lift the ball-check off its seat. As the piston moves the other direction the fuel would attempt to drain back down the tube but it is blocked by the ball-check valve.

Once you've removed the carburetor from the tank, inspect and clean the intake screen and be certain the ball check is free and moves when you shake or invert the carburetor. Carburetor cleaner will usually clean up the screen and free the valve, but care must be used with Nylon pipes since some cleaners may harm the plastic.

Nylon pipes are easily removed with a small socket wrench since they are threaded into the carburetor body. To remove a brass pipe, clamp the pipe in vise jaws with no more force than necessary

Tank-top carburetors may have either Nylon or brass feed tubes. Nylon tubes screw into the body; brass ones are pressed in.

and gently pry upwards against the carburetor body with two screwdrivers to withdraw it. To replace a brass pipe, use a vise to press it in until it projects 2⁹/₃₂ inches from the carburetor mounting surface. Other service is routine, such as cleaning and inspecting the adjustable needle valve, checking the throttle and choke shafts for wear, and being sure that the mounting surface is clean and flat and the gasket is well sealed at the mounting flange.

Pulse-lift carburetors. Somewhat more sophisticated than suction-lift carburetors, the pulse-lift type has many variations. But if you understand the operating principles they are not difficult to service. Once a pulse-lift carburetor is removed from the tank, you can recognize it by its two fuel pipes, one longer than the other. These pipes both have screens on the ends but do not contain ball checks; shaking the carburetor should not produce an audible rattle. The check function is provided by diaphragm-type flapper valves.

While suction-lift carburetors store fuel above the ball-check valve, in the pulse-lift units, negative pressure causes the fuel to rise up in the longer pipe and through the diaphragm flap valve. This also compresses a small spring above the diaphragm. As the negative pressure is reduced, the spring pushes the diaphragm down and closes the flap leading to the long pipe. A second valve then opens and the fuel is forced by the diaphragm spring and manifold pressure into a small reservoir in the top of the tank. The short pipe extends into this reservoir and from here the fuel is transfered in the usual manner to the carburetor throat via the adjustable needle valve passage.

One common variation of this carburetor is the diaphragm-actuated automatic-choke variety discussed earlier in this chapter. Remember, even though the choke function and pumping function use different areas of the same diaphragm and carburetor body, they are quite separate otherwise and you should think of them that way.

Other carburetors operating on the pulse principle use small pumping diaphragms about the size of a postage stamp. These are contained under a rectangular cover on the side or top of the carburetor body and can be replaced without removing the carburetor from the engine. This entails some extra care since it is hard to hold the pump spring and its seat in place unless the engine is tipped so they stay there by their own weight.

Since there are no check balls in the pipe, cleaning

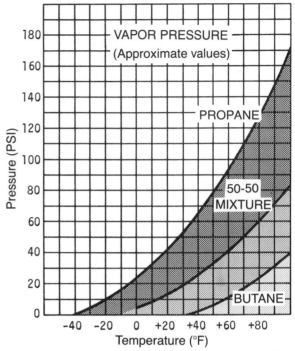

Propane and butane have very different vaporization characteristics. Propane vaporizes at -44° F, while butane does not vaporize below 32° F. Also, at 120° propane tank pressures may reach 225 psi.

the screens should be sufficient. If the screens or pipes are hopelessly plugged, the pipes can be replaced following the procedure on the facing page.

GASEOUS FUEL SYSTEMS

Today, with recreational vehicles using liquified petroleum gas for cooking, and camping lodges and the like storing LPG on-site, the use of propane, butane, and even natural gas to power small engines becomes very attractive. With LPG, a supply of fuel can be stored indefinitely for operating a generator, water pump, or other equipment. Unlike gasoline, LPG does not deteriorate in storage and is not easily pilfered.

Natural Gas

Natural gas is an ideal motor fuel if it has been scrubbed of pipeline debris and contaminants such

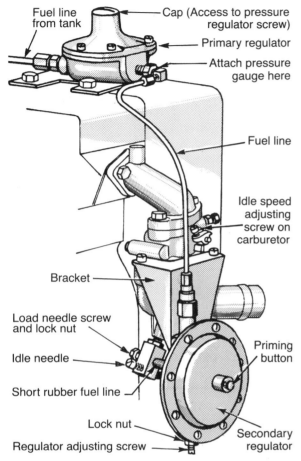

Typical gaseous-fuel installation shows primary regulator on top. A fuel line runs to the secondary regulator, which is closely coupled to the carburetor.

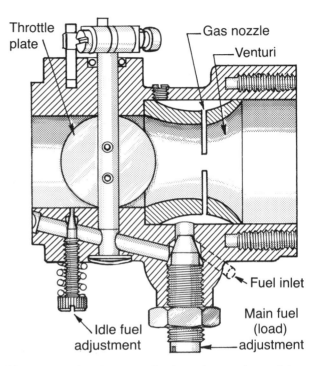

Most gas carburetors are simple and rugged, consisting of a body, throttle plate, and a load adjusting screw.

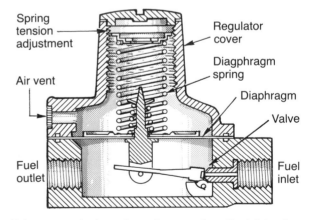

Primary gas-fuel regulator. In operation, the inlet valve is opened as the diaphragm moves down in response to fuel demand. Incoming fuel raises the diaphragm against atmospheric pressure and regulator spring. Fuel inlet valve establishes a balance to maintain desired fuel flow and pressure.

as sulphur and other random products. Unless you have your own oil or gas well—and a surprising number of people do—that is no problem since commercial natural gas is a carefully controlled product.

Natural gas has an extremely high octane, or antiknock, rating, as high or higher than aviation gasoline. Thus, it is not likely to detonate in your engine. And since it is a "dry" gas and arrives through the pipeline to your engine in an already vaporized form, carburetion is relatively simple and there is

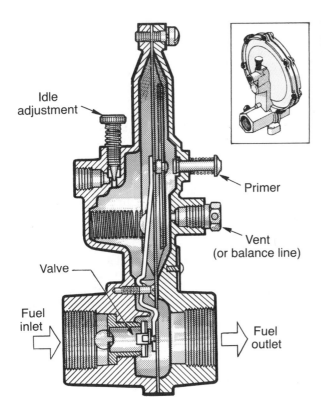

Idle adjustment

Primer

Vent (or balance line)

Valve

Fuel inlet

Fuel outlet

Secondary gas-fuel regulator. When the engine is not running, diaphragm spring pressure holds the fuel-inlet valve closed. As the engine starts, negative pressure from the carburetor moves the diaphragm which then opens the valve.

compensate for the loss of tetraethyl lead's lubricating effect.

Propane and Butane

Much of what I've said about natural gas also applies to propane and propane/butane mixes. They are both of relatively high octane, and engine oil remains clean for extended periods. There are, however, some significant differences between natural gas and these LPG fuels.

As the name liquified petroleum gas indicates, both propane and butane must be stored as liquids in government-approved tanks or pressure vessels. At normal atmospheric temperatures and pressures they vaporize instantly. The design and construction of the tanks is closely regulated as with welding-gas tanks. That alone should advise you that some precautions are necessary.

Propane differs only slightly from butane in heating value. Propane, however, is generally preferred as an engine fuel because of its ready vaporizing characteristics at low temperatures. Even at temperatures well below 0°F, propane vapor will still feed from the tank. This contrasts with butane, which does not vaporize below 32°F. Butane has also been found to vaporize poorly even at 40°F.

On the other hand, a tank of propane, perhaps exposed to the sun, may have a pressure as high as 225 psi at 125°F. Under the same conditions, butane

practically no dilution of the oil because there's no liquid fuel to wash down the cylinder walls.

Natural gas is an ideal fuel for an emergency generator or an irrigation pump engine, but it should be installed professionally by a service agency thoroughly familiar with the safety and technical requirements. An engine, for example, may require a somewhat higher delivery pressure and available volume than a gas furnace or stove. You should also consult the engine manufacturer about the type of valves and seats in your engine. For example, as a dry gas it lacks the natural lubricity of gasoline. But with the advent of unleaded gasoline, most engines are capable of operating on gas. This is because engine makers are using tougher valves and seats to

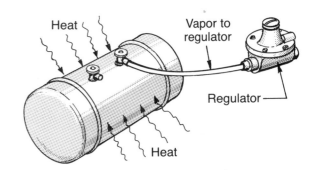

Heat

Vapor to regulator

Regulator

Heat

Stored as a liquid under pressure, LPG must have heat to convert to vapor. On water-cooled engines, heat is added as hot water passes through a vaporizer. Air-cooled engines must rely on heat from the air surrounding the tank.

will have only 56 psi. Thus, vendors of LPG commonly blend these gases to stabilize tank pressures over winter and summer months.

Pressure regulation. Since the pressure at which LPG is delivered from the tank can range from a possible several hundred pounds per square inch to maybe 10 psi in extreme cold, it is essential to have a gas regulator that will reduce and stabilize this pressure at about 4 to 6 ounces per square inch. This is the job of the primary regulator. It is much the same as the regulator outside your home if you have gas for cooking or heating.

To supply the fuel to the engine on demand a secondary regulator is also needed. This acts somewhat like a float in a gasoline carburetor. When the engine is not running the regulator must close off all fuel flow to prevent a leak. As was pointed out for gravity-feed float bowls, the regulator should never be totally depended on and some form of positive shutoff should also be in the system.

Both primary and secondary regulators, working together, respond to the negative pressure in the carburetor throat and control the fuel/air ratio. In many systems the two are combined in a single unit called a *two-stage regulator.* Do *not* attempt to repair or service gas regulation equipment. This is a specialized field requiring professional equipment and training.

Gaseous-Fuel Disadvantages

Gaseous fuels are not ideally suited to small, mobile equipment such as mowers and tillers. Liquid petroleum gas, in particular, requires heat to change into a vapor. If you were using it to fuel a large, water-cooled car or truck engine, you would probably withdraw liquid from the tank and then vaporize it by passing it through a series of coils heated by water from the engine's cooling system. With small air-cooled engines, you'd have to rely on the air surrounding the tank for the heat needed to turn the LPG to vapor—which is the way it would come out of the tank. The small fuel tanks practicable for use in small equipment would quickly chill and inhibit the flow of vapor, especially with butane. Moreover, gaseous fuels are heavier than air and, if there is a leak, can collect in a low spot where they would pose a fire hazard. Thus, it is never a good practice to operate a gaseous-fueled engine in a closed building where free ventilation is not available.

In addition, although personal repairs to a gasoline-delivery system are well within most engine owner's skills, the repair of gas equipment is better and more safely left to professionals. Gas systems, however, require little service since they are not subject to gumming and other problems associated with gasoline. About the only service you may want to attempt yourself is to use a small brush and soapy water to check the connections and joints for leakage now and then. There must be absolutely no bubbles indicating leaks. Never try to hook up gas carburetion with the type of flexible tubing and fittings sometimes used for domestic heating appliances. Engine vibration can cause quick and complete failure of such hardware. Again, I emphasize that LPG is often an ideal fuel, but servicing the components is strictly for the professionals.

8

Governors

One of the more baffling devices on small gas engines is the governor. Even experienced mechanics find the little arms, springs, and linkages that connect to the governor and carburetor confusing. Yet a basic understanding of how a governor works will help make service and adjustment a routine job, even though they have varying formats and hookups.

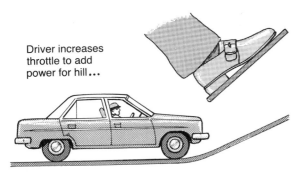

Driver increases throttle to add power for hill...

In your car, you can manage the throttle and the power to suit conditions as you see them. A portable or emergency generator, below, has no way to anticipate a remote load applied instantly.

But generator cannot "see" that switch will be closed here

When driving your car, if you sense a loss of speed because of a hill you automatically tread harder on the accelerator pedal. Or, while sawing firewood, you position your saw with the engine at idle and then squeeze the throttle to make the cut. Now consider an emergency generator, or the portable one you're using to feed a saw, drill, or other tool. There is no way either can anticipate that you will throw in the load switch or that an automatic switch on a refrigerator or well pump will cut in. Even though the engine is running at a speed that will produce 60 cycles per second (60 Hz) without a load, the throttle is nearly closed and it is putting out only enough power to overcome its own friction. Suddenly, the load goes from no-load to full-load. Unless the throttle opens very quickly the engine will slow down and perhaps die.

THE GOVERNOR'S JOB

The governor's job is to sense a slowing from load and immediately open the throttle just enough to match the power required by the load. This same basic throttle adjustment takes place thousands of times when you mow your lawn, till your garden, or move snow. Your mower encounters thick grass, thin grass, and if you tilt it back to turn it around, no grass at all. Yet the engine speed remains just about the same. Your tiller or snow thrower may also hit dense going, but the governor reacts and prevents the engine from stalling. The governor also does something else very important: It prevents the engine from overrevving.

Even if you could keep one hand on the throttle at all times you could not make these adjustments quickly or accurately enough. If you doubt this, try it sometime by disconnecting the link from the governor to the throttle and fixing a wire control to the

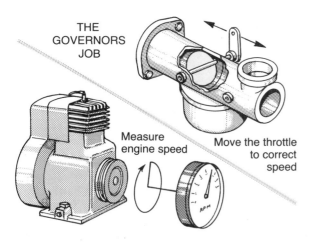

THE GOVERNORS JOB

Measure engine speed

Move the throttle to correct speed

The governor must constantly monitor crankshaft revolutions and sense increases or decreases in speed. It must then move the throttle plate as required, the equivalent of your foot on the gas pedal when driving.

point where you can reach it while working. Be careful! Your engine can come apart in a hurry.

HOW THE GOVERNOR WORKS

If by now you've concluded that a governor is essentially a substitute for a human hand or foot on the throttle, you're right. But a governor's first function is to measure rpm. Secondly, it must physically move the throttle. Almost without exception in small gas engines, this movement comes from the centrifugal force of small weights or from air pressure against a vane-like member under the air shroud.

Air Vanes and Weights

Air-vane governors consist of a thin metal or plastic blade mounted so it's free to pivot in the air

How An Air-Vane Governor Works

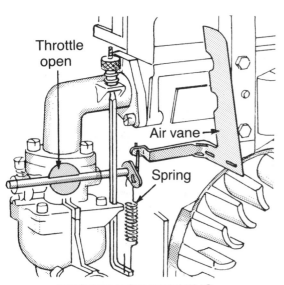

Throttle open

Air vane

Spring

ENGINE NOT RUNNING

1. When the engine isn't running, the spring holds the throttle plate in the open position. This would be the normal starting position for the air vane.

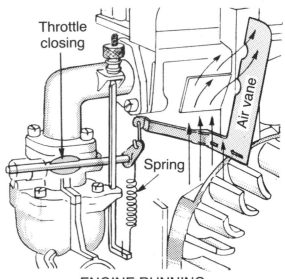

Throttle closing

Air vane

Spring

ENGINE RUNNING

2. When the engine starts, air from the flywheel blower moves the vane and overcomes spring tension until the balance point is reached.

stream developed by the blower blades on the flywheel. The faster the engine runs, the greater the air pressure against the vane. Centrifugal-weight governors use small, pivoted weights spinning on a shaft inside the engine and driven off the crankshaft or camshaft gear. The faster the engine runs, the farther the weights extend on their pivots.

The first point to remember—no matter which type of governor you're working on—is that the vane or weights can deliver a force only when moving outward in response to an increase in engine speed. The weight or air-vane force is always trying to close the throttle and slow the engine down.

That means we must have a second force to oppose or balance the weights or air vane. This normally takes the form of a spring hooked to the governor arm. The second point to remember, then, is that the spring is always trying to pull the throttle open and increase speed.

With the engine on your mower or tractor stopped and the manual throttle in the "Run" position, take a look at the throttle arm on the carburetor. If you want to be sure of what you're looking at, remove the air cleaner and with the choke wide open look into the carburetor throat. You'll see that the throttle plate is on edge, or close to it. If you were to start the engine and block the throttle in this position with no load, the engine would overrev quickly and probably dangerously. But as soon as the engine starts, centrifugal force acts on the weights, or air on the vane, and the spring that was holding the throttle open is now opposing another force trying to close the throttle. When the two forces reach a point of balance, the throttle plate will assume a partially open position and stabilize the engine speed at maximum governed no-load speed.

As soon as you load the engine and it starts to slow down, the weights lose some of their centrifugal power, or the air pressure against the vane is reduced. Now the spring has the greater power, and it immediately acts to pull the throttle open and speed the engine up until the forces again balance at or near the original speed setting.

Choosing the right weights. A manufacturer will design and install weights with certain characteristics to provide a desired governor performance. In many cases these weights will be aimed at providing satisfactory performance in most general-purpose applications. The same applies to air-vane governors. But in some cases an engine may be intended to drive a specific piece of equipment with somewhat

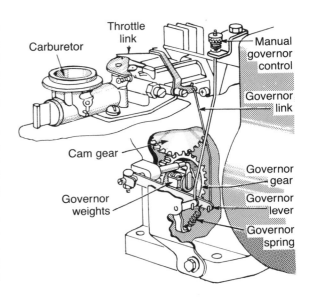

This centrifugal governor uses a spring to provide a balancing force, as in the air-vane governor, but here the spring is opposed by the centrifugal force from small governor weights.

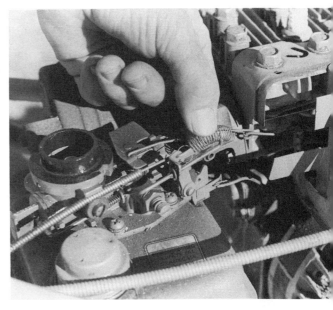

Typical spring for an air-vane governor. Spring tension is increased as you move the throttle control open.

Typical centrifugal governor shows two small weights free to swing under centrifugal force. In operation, the weights shift the governor spool out against the internal governor arm.

different governing needs. Thus, if you are rebuilding your engine or have several engines of a common type, you cannot take it for granted that all weights are the same. *Check the parts list for your engine and application.*

Governor Springs

Often it is easier and cheaper for engine manufacturers to provide several different governor springs for a series of engines. You may find these in the parts list distinguished by color—for example, cadmium-plated, black, red-and-white, green, and so on. The difference in these springs is their *rate,* or amount of tension they will exert when stretched a given distance. One spring of a given size may be quite stiff and have perhaps 2 pounds of tension when stretched half an inch, while another may have only 1 pound. The soft spring may allow the engine speed to sag severely under load and then return to

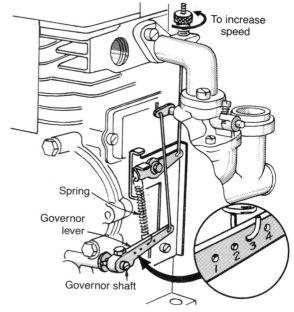

In general, moving the spring so tension is increased and location is shorter from pivot to spring will increase sensitivity. Always reset top speed to stay within maximum speed recommendations after making any governor adjustments.

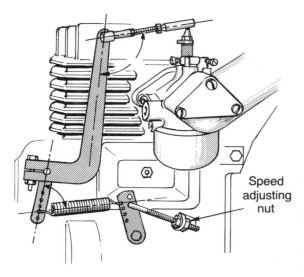

You probably need never change linkage-arm angles unless you are trying to fine-tune a generator engine.

This threaded link allows for fine tuning of the governor response on a generator engine.

governed speed only gradually. The stiffer spring may give a snappy action and, in effect, "jump on the load."

Linkages

Almost all small engines have a governor arm with a series of holes into which you can hook the spring. This is one more provision for tuning gover-

nor response to the performance demands on the engine.

Before unhooking the spring and linkage, always make a detailed sketch of the exact connections and which links go in which holes on the carburetor and control arms. In general, the closer the spring or link is hooked to the pivot point, the faster the governor action. If you have an engine that seems a little slow on response, you might want to gently unhook the

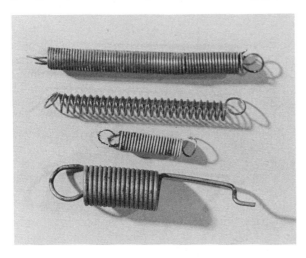

Governor springs come in variety of shapes, sizes, and lengths. Each is carefully worked out for a specific engine and load.

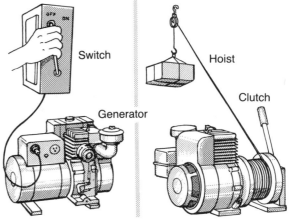

Throttle response to speed change must be extremely quick in some applications, while in others a more leisurely response is required.

The internal governor arm—pointed out above—rides against the weight spool. It must work freely and without binding if the machine is to operate properly and perform the work for which it was intended.

spring and move it one or two holes closer to the pivot. If this induces surging and overspeed when the load is released, you've gone too far.

Linkage angles. Ideally, a governor should respond with a modulated reflex closely akin to human movements. It should open the throttle quickly in response to load and then, as the engine speed approaches the desired point, ease off the rate of throttle movement so the engine does not overrun. This is exactly the way you handle the throttle in your car or boat.

Note that as the linkage and governor arm and throttle are positioned at about 90 degrees, the throttle movement for a given governor-arm motion will be maximum. As the lever swings past center and the angularity increases, the governor arm continues to move but the rate of throttle movement slows because of the geometry involved. Some high quality engines have linkages with threaded-on terminal ends so you can change the length of the link. Making such adjustments is usually a fairly technical business, however, and unless you have a genuine need for such governor finesse, stick to the original settings.

Load droop. Most small-engine governors are constant-speed types. You start the engine on your mower or snow thrower and go to work. In the case of a rotary mower, the governed maximum speed has been set to hold the tip speed of the blade, running no-load on dry pavement, at no more than 19,000 feet per minute. Yet no governor of the simple types used on small engines can maintain rpm perfectly.

If you start your mower on a walk or drive and set the throttle at "Run," there is no load and the engine is said to be running at high-idle. With a

Governor arm

Governor shaft

In engines where the rotating weights and spool are mounted on a shaft seated in the crankcase cover, hold the governor arm as shown as you slip the cover in place.

20-inch blade your engine will be turning about 3,600 rpm. As you move off into heavy grass you'll hear the engine slow down slightly as it is loaded. The governor will open the throttle and if the engine can easily handle the load it will probably return close to the governed speed. If the engine does not have enough power to handle the load at 3,600 rpm it will slow down, the throttle will be wide open, and the engine will basically be overloaded. In most small equipment this is a frequent, transient state and is not especially serious unless the engine lugs for extended periods at much lower than designed speed.

On a generator set, speed droop under load is of greater concern. When the engine slows down it is not maintaining the 60 Hz (cycles per second) most equipment you'll power is designed for. The output voltage will usually drop as well. This means that it is customary to set the governor so rpm is slightly high at no-load and the frequency is about 62 or 63 Hz. Under load the engine should settle down to about 60 Hz.

Though this rather crude level of frequency control is satisfactory for running power tools or keeping your refrigerator or furnace blower going, it won't keep an electric clock or timer accurate. And if there is a momentary overspeed when part of the load drops off the line, the power surge could be harmful to a TV, home computer, or similar device. That is why small-engine governors are only marginally suitable for generators.

Governor sensitivity. Not all jobs require, or can tolerate, the same degree of governor sensitivity in response to load. For generators, the ideal is an instantaneous reaction to load that would open or close the throttle, from no-load to full-load or vice versa, in a fraction of a second when you throw the switch. But that type of response would be extremely dangerous in some other small-engine applications, such as those powering a hoist, backhoe, or bucket loader. Here, an instant response would be the equivalent of kicking the throttle quickly when letting in the clutch on a stick-shift car. Engine surging or hunting is one common result of an oversensitive governor. In most cases, sensitivity can be adjusted to suit the equipment.

SERVICING WEIGHT-TYPE GOVERNORS

In most such governors, the motion of the weights is applied to a sliding, often spool-shaped member. As this spool slides back and forth a lever arm rides

The external governor-arm setting must be reset relative to the internal weights. A manual can tell you whether the shaft extension, being turned and held here by pliers, should go clockwise or counterclockwise. Throttle is held open during adjustment.

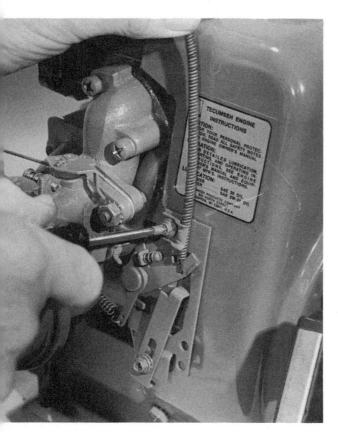

If necessary, relocate the Bowden control wire for full movement and tighten the clamp securely as shown.

on it and communicates the motion to the external governor arm outside the crankcase. If the rotating weights and spool are mounted on a shaft seated in the crankcase cover, and the cover is removed, be sure the arm extending into the crankcase is aligned with the engagement slot or other detail of the spool when reinstalling the cover. The following alignment procedure works best for many small engines:

1. Place the engine on its base so that the arm hangs down vertically, then align the arm and spool as the cover is slipped into place again.
2. Before closing down the cover plate and tightening the screws, peer into the crankcase with a flashlight to see if the parts are correctly mated.
3. With the plate secured, be sure the linkage moves freely and that there is no hint of interference.

In other engines, access to the weight assembly is best gained by removing the bottom pan or plate of the engine. Since the governor weights, shaft, and spool are not heavily loaded parts or subjected to heat, it is unlikely that major service will be needed. Whenever the engine is open, however, these parts should be inspected.

Inspecting Weight-Type Governors

Begin by making sure that the support shaft for the weights and spool is firmly pressed into the crankcase and not loose. Examine the running surfaces for scratches, wear, or indications of lack of lubrication. The weight assembly must spin freely on the shaft without binding or excessive looseness. Check the weight pivot pins for security and free-swinging weights. At the same time, rotate the weight drive gear to spot any broken or chipped teeth. If you find any, you'll probably have to replace the mating cam gear. Finally, be sure that the contact surfaces of the governor internal arm and the weight assembly are clean and smooth. Unless the engine has run without oil or has had sand or foreign material circulating in it, you are not likely to find serious problems here.

Governor Adjustment

Unless your engine's governor is not meeting specific governing requirements, or the governor or linkage has been disassembled, you'll seldom need to adjust the governor in normal service. In those rare instances, however, you'll need some form of tachometer to adjust engine speed properly.

Governor-arm position. On the flyweight-type governor with weights inside the engine and the control arm outside, check a shop manual for the proper

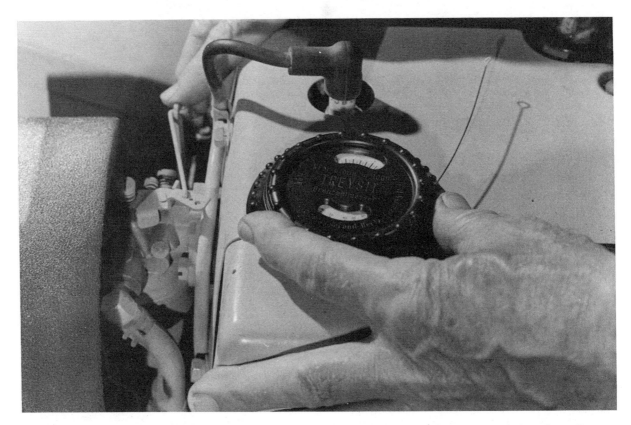

To use a vibrating-reed tachometer, place it on a flat surface and slowly turn the dial until the small wire suddenly vibrates vigorously. The dial reading tells you the rpm.

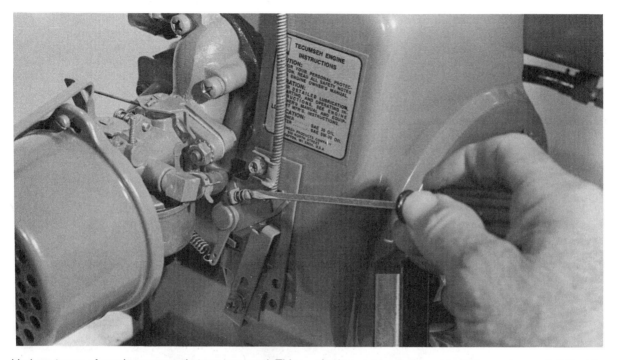

Various types of maximum-speed stops are used. This one has a screw that blocks travel of the spring-tensioning arm at the selected maximum full-throttle speed for proper operation.

positioning of the external governor arm on the extension of the internal arm shaft. Although there are many detail variations, here are the basic principles.

With the engine stopped, there is no centrifugal action of the weights and they are simply hanging free on their pivots. What you want to do is to force the spool back firmly into contact with the weights and establish this position as a starting point for adjustment. The external governor arm is usually clamped with a light screw clamp to the internal governor extension shaft. By loosening the clamp you can usually swing the external arm. That enables you to turn the extension shaft with pliers, or a screwdriver if it has a slot, so as to swing the inner arm snug against the weight spool. In many horizontal-shaft engines this means rotating the shaft to the left and in some vertical shaft engines it means turning it to the right.

The governor spring will probably be holding the throttle open or it will do so if you move the control lever to the *Run* position.

1. Hold the throttle plate open and turn the extension shaft back the opposite direction.
2. Retighten the clamp screw. The movement of the weights should now have their full range of travel available to close the throttle in opposition to the spring.
3. Start the engine to be sure, but be ready to stop it immediately if the engine speed increases rapidly. Follow the action of the governor arm and throttle during several accelerations.

Throttle control. The next thing to check is the throttle linkage control. This applies to all engines with a remote throttle control up on the mower handle or on a tractor panel. Most such controls are merely Bowden wires sliding in a wire-coil casing, with a lever or knob at the handle end.

If an engine doesn't seem to run at full speed—and the throttle-wire connection has recently been loosened or removed for service—examine the small, screw-retained clamp that locates and secures the end of the Bowden wire casing to the engine in the carburetor area. If the casing has slipped or been clamped so that its end is too near the governor connection, the full governor-spring tension cannot be applied and the engine will run slowly.

Often a movable plate or tab, or stop screws, limit the distance the Bowden wire can pull the control. Try moving the throttle at the control handle and see if the engine end of the wire actually pulls the movable part to which the governor spring attaches against the stop. If not, re-clamp the Bowden wire casing a bit farther out until you get full travel.

Engine speed. To make the final speed adjustment, you will have to work with a running engine.

Begin by starting the engine and letting it warm up at medium speed. While it's warming up, take a long, hard look for any hot parts or rotating parts that might entangle you. If you're working on a rotary mower, you must always expect the blade to be turning even though you may have a blade/brake clutch. Make sure you have no loose or dangling clothing. Most important, make no quick, spontaneous moves. *If you are going to place a tachometer or adjust a screw, think through how you'll position your body, your feet, and both hands.*

1. Once the engine is warmed up, make sure there are no loads on it.
2. Take a tachometer reading of the maximum full no-load speed. Remember—to increase speed you must increase governor-spring tension; to decrease speed, ease off spring tension.
3. If the speed is too low, you may find a screw or lock tab that limits either governor spring tension or throttle travel. Use your finger or the screwdriver tip to push the throttle towards the *open* position.
4. If the throttle moves easily and the engine speeds up, the throttle is not being blocked and spring tension is too low. Adjust the screw or tab to allow more tension. If you cannot move the throttle, look for a screw or tab that is holding the throttle shut and adjust that.

If there is no screw thread or movable tab adjustment, note whether the governor spring is anchored at one end by a tang on a sheet-metal throttle arm. Such spring and tang hookups are usually adjusted by bending the tang slightly to increase or decrease spring tension. If the tang will be relatively inaccessible, as it is in most cases, try sawing a 1/16-inch-wide slot in an L-shaped piece of quarter-inch steel rod to slip over the tang and tweak it a small amount.

SERVICING VANE-TYPE GOVERNORS

Because of the inherent simplicity of vane-type governors, there's really not much in the way of service that can be performed, save for replacement. What you can do is to make sure that the vane or linkage is not distorted, and that the vane is not binding in its pivot.

This is repetition, but I'll say it again so you remember it: *The weights or vane always want to slow the engine down; the spring always wants to speed it up.* Follow the two forces through to see how they affect the throttle position, and you can figure out how to adjust almost any governor.

Generator Governors

Since generators are used to drive electric motors and sometimes electronic devices, the generator output must closely approximate utility power. This means that the alternating frequency must be quite close to 60 Hz (50 Hz in some countries) and the voltage must be as close as possible to the power line's 115 to 120 or 230 to 240 volts.

Your first concern is that the engine powering the

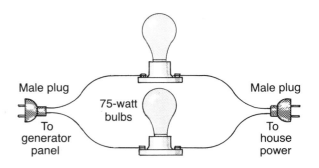

A simple frequency checker for adjusting your home generator requires two 75-watt bulbs, a length of wire, two male plugs, and two bulb sockets.

generator has enough horsepower at the rotating speed required for 60 Hz to carry motor-starting and other loads, which are substantially higher than running loads. The generator, too, must also be rated for and capable of supplying that power. The rule of thumb is that you must have 2 horsepower per kilowatt. Thus, a 3,500-watt (3.5 kilowatt) generator needs at least a 7-hp engine; a 5,000W (5kw) generator requires at least 10 hp. A 3.5kw generator is just about the smallest practical size for emergency power.

Your second concern—and from the standpoint of this chapter, your most important one—is the governor's ability to respond to those loads.

Testing a generator governor. Let's assume you've chosen the right engine and generator for your needs, and have provided the proper transfer switch so that no power can feed back into the power company's lines. Before throwing the switch to feed power into your household circuits, you must first cut off all circuits, including lights, TV, refrigerator, air conditioner—everything.

Probably a well pump, if you have one, is the greatest single motor-starting load your generator will be called upon to handle. These motors normally range from $\frac{1}{2}$ to 1 hp, but the starting current draw is much higher than the running draw.

1. Using a well pump, be certain its household fuse or circuit breaker is closed. Since you'll probably be testing outside to avoid exhaust fumes, open a garden hose and wait for the pump to cut in.

2. When it does, the engine will suddenly start to pull hard and the governor will act to open the throttle. If you have a frequency meter plugged in, see how many cycles the speed drops during this motor startup phase.

3. Lacking a frequency meter, a 50w light bulb in the circuit or plugged into another generator outlet will do. If it dims slightly, then returns to normal brightness in about a second, your unit is doing fine. If it dims and stays dim, and the generator continues to labor, try listening to the well-pump motor.

The latter symptoms—particularly a labored sound from the pump motor—tell you the motor is trying to start on the starting coils but is not getting up to speed and will blow the circuit breaker in a moment.

Be ready to actuate the carburetor butterfly lever with your finger or a stick. If you can open it a trifle, and the generator engine speeds up allowing the pump motor to run normally, you know the governor is simply not opening the throttle enough to provide full power. If, however, the throttle is wide-open and the engine is still laboring, the engine just isn't powerful enough for the job. *Open the circuit-breaker switch before you damage the motor or generator by overloading it.*

If, in the preceding test, the governor simply wasn't opening the throttle enough to pick up the load and get the engine back up to speed after a momentary speed drop, you may have a governor intended for other than generator use. You can also try moving the governor linkage connection one or two holes closer to the pivot or changing to a slightly stiffer spring rate. If that doesn't help, write to the

In many cases, a well pump is the heaviest and most critical load; by squirting the garden hose until the pump cuts on you can watch the frequency meter and observe what happens. You should expect to see the frequency drop a few cycles momentarily and promptly return to 60 Hz.

service department for the manufacturer of that particular engine, giving all details including engine model, specification number, and serial number, as well as the make and type of generator.

How to get 60 Hz. Suppose your engine, generator unit, and governor perform well; all you did was clean the carburetor or perform some other minor service that involved the governor and throttle parts. How can you be certain, after reassembly, that the governor setting still allows 60 Hz? A frequency meter will tell you, but lacking that there is an easy way to check using two 75w bulbs, one spliced into each side of an extension cord that has been split lengthwise and given two male ends. After starting your generator engine and warming it up:

1. Plug one end of the cord into the generator. Be extremely careful of the open prongs on the other end—they're live with electricity!
2. Plug the other end into any household outlet. The bulbs will be dimmer than normal because they are at half power. They may flicker rapidly—showing that your governor setting is way off—or they may slowly brighten and darken, which means you're quite close to 60 Hz. The power passing through the bulbs is doing so because the utility and generator frequencies aren't the same. Power is more or less spilling back and forth between the bulbs.
3. To adjust the governor, try moving the throttle slowly and in small increments to see which way to change speed to bring your engine into synchronism.
4. If tweaking the throttle open slightly causes the bulbs to darken and stay dark for a second or two, tighten the governor spring slightly to increase engine speed. If closing the throttle slightly darkens the bulbs, slacken off a bit on the spring.

Your engine will drift in and out of synchronism, but you should reduce bulb flickering to a minimum. Your engine is now running at 60 Hz speed but at no load. You can either tighten the spring slightly to boost the speed to about 62–63 Hz, or you can plug a load such as a shop motor into another generator outlet and let it run while you again adjust the governor spring for minimum brightening and darkening of the bulbs. Now, when you cut the load the engine will speed up slightly but that's your best setting.

As I said before, no governor of this type is perfect, but you should come close.

9

Flywheels and Flywheel Brakes

A flywheel's primary job is to store energy to carry the crankshaft around during the three non-power strokes of a four-stroke engine and the combined intake and compression stroke of a two-stroke. Yet that is just one of many flywheel functions on today's typical small gas engine.

WHAT THE FLYWHEEL DOES

Nearly all small engines also use the flywheel as a source of electric power for ignition, which comes from two or more magnets in the flywheel rim. These magnets may be made of high-grade magnet

Most modern flywheels have magnets inside or outside the rim for ignition and often for alternator power.

steel or a ferric material with superior magnetic properties. Because the ferric material is not well suited to bouncing on a concrete floor, be sure to inspect the flywheel carefully should you drop it.

The development of the alternator and the solid-state diode made it practical to further use the flywheel as a battery charging source for direct current. And when automotive-type electric starters replaced the cumbersome starter/generator, the ring gear on which the starter motor engages soon became another part of the flywheel.

Finally, the flywheel is used to blow the cooling air through these air-cooled engines. In most cases the fins are simply cast with the flywheel, but in some cases they are of plastic material fixed to the wheel rim.

Basic Flywheel Requirements

To perform its many jobs adequately, the flywheel on your small gas engine must meet certain requirements:

- It must be closely balanced to run smoothly.
- It must be extremely durable and free of cracks and flaws to withstand the high centrifugal forces without exploding.
- It must be securely attached in perfect alignment with the crankshaft to run true without shifting position.

Flywheel Brakes

Walk-behind rotary mower engines were profoundly affected by the Consumer Product Safety Commission standard of July 1982. The ruling mandated that the mower operator maintain a grip on a bail on the mower handle at all times while starting and mowing. If the bail were released, the engine and blade had to come to a complete stop within

If your engine has an automotive-type starter, the fly-wheel will have a ring gear into which the starter pinion-gear moves. When you press the switch, a solenoid engages the starter gear with the ring gear.

three seconds. Whether the starter were electric or a rope and handle relocated near the operator position, the idea was to keep the operator's hands and feet well away from the cutter blade. Exceptions to this were mowers with blade-brake/clutch systems that stopped the blade but not the engine—covered on page 300.

Most mower builders opted for the stop-and-restart system, which meant starters, ring gears, and alternator chargers were now essential to mower owners who preferred not to pull a rope on every restart.

The safety regulations also meant the flywheel now had to provide a braking surface to stop the engine. Briggs & Stratton uses a steel band lined with brake-lining friction material that wraps almost entirely around the rim of the flywheel and clamps down to act as a drum brake. Tecumseh uses a spot-type brake with a lever action to press a pad of brake material against the flywheel. The point of pressure may be either the bottom edge of the flywheel or the inner surface of the flywheel rim.

FLYWHEEL REMOVAL AND SERVICE

At one time, removing the flywheel to service or replace the ignition and condenser beneath it was a frequent procedure. With the advent of electronic ignition, flywheel removal is far less common. But you'll still have to do it for most major repair work, to replace seals, or to see if a starting problem is the result of the flywheel having shifted on the crank-

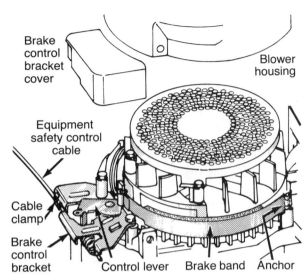

Brake control bracket cover

Blower housing

Equipment safety control cable

Cable clamp

Brake control bracket

Control lever   Brake band   Anchor

Band-type flywheel brake, above, grips the rim of the flywheel with a metal brake band lined with friction material. The pad-type brake below acts much like a brake pad on a car and bears against the flywheel rim to halt rotation. There's also a starter interlock switch since this is an electric-start engine.

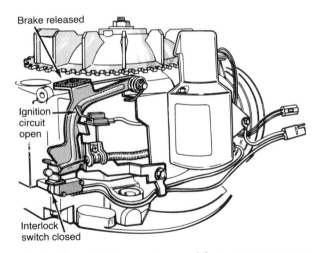

Brake released

Ignition circuit open

Interlock switch closed

shaft after striking a stump with a rotary mower blade. The latter condition will throw the ignition timing off since the magnets in the flywheel will be disoriented with respect to crank and piston position. Checking the alternator may also require flywheel removal.

Flywheels are retained on the tapered crankshaft by either a nut and a spring washer, or, with some rope starters, by a threaded starter clutch that serves as a nut. Either retaining setup can be loosened by following the procedure on page 59 for removing the starter ball clutch. With the nut or clutch removed:

1. Release the flywheel brake, if your machine has one, and keep it released. Wire the operator control in the *Run* position if necessary.
2. Use a puller if the flywheel has two or three threaded puller holes near the hub. Be careful of the crankshaft threads, and leave the nut on the end of the shaft to protect it. Use the centering button on your puller if it is so equipped. *Do not use jaw-type pullers under the flywheel rim; this can cause serious damage and distortion.*
3. Some flywheels have puller holes but no threads. There's a factory tool for these, though you can also tap for threads. Flywheels without puller holes are generally intended to be removed with knock-off tools.
4. In some worst-case situations, even strong tension on the puller screws or repeated attempts with a knock-off tool won't budge a tightly seated wheel. Here, you might try heating around the flywheel hub with a propane torch. *Be careful!* Make sure that no gasoline or gasoline fumes are present, that the carb and fuel tank are removed from the machine, and that you are working outdoors. Also be sure no heat gets near alternators, solid-state ignition components, or other electrical parts. The heat must be confined to the flywheel hub.

Flywheel Inspection

Since the flywheel is subject to such high stresses and can be dangerous if it disintegrates, you should always inspect it carefully whenever you remove it. The most dangerous condition is cracking. Small cracks may not be visible to the eye even with magnification. To check for them:

1. Suspend the flywheel by a piece of string tied through the center crankshaft hole.

2. Strike the wheel rim lightly with a small hammer or tool handle; it should ring with a bell-like tone much like that from a fine piece of chinaware.

Another way to check a suspect flywheel is to brush it with kerosene, wipe the surface completely dry, then coat it thoroughly with chalk. Place it in a warm spot overnight and examine it the next day for lines or spots where kerosene has seeped out and discolored the chalk.

Causes of flywheel cracking. The two common causes of flywheel cracking are striking an object with a mower blade and a loose flywheel. In both

Take extra care when removing or installing a flywheel to be certain it is free of cracks, that the key is correct and properly seated, and that the nut is fully torqued.

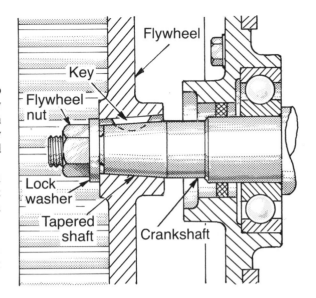

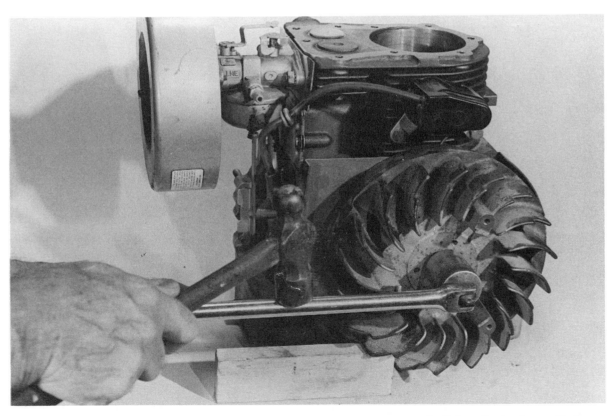

A block of wood, ½-inch drive socket, and a sharp rap with a hammer should break loose the nut on heavy cast-iron flywheels. On lighter flywheels, you'll almost certainly break the blades with this method.

This flywheel has two puller holes. The puller is homemade and is heavier than it needs to be.

cases, expect to find the flywheel key at least partially sheared and the keyway battered. Note that most engines do not use steel keys in the flywheel and crankshaft. A softer metal such as zinc is used, and you must never replace a soft key with a steel one. Remember, too, that keys may vary depending upon the type of ignition system used. Since they may be of different shapes and materials, the correct part number is important.

You can easily differentiate flywheel cracking caused by striking an object, as opposed to a loose flywheel, by the location of the crack. If looking down into the rear of the flywheel with the keyway at the 12-o'clock position, away from you, the crack is at the right side of the keyway, the cause was striking an object. If the crack is to the left side of the keyway it was caused by a loose wheel. Keyway

battering in the latter case comes from the power impulses of the piston.

Left-side cracking may also be associated with fretting corrosion as evidenced by dry, rust-like deposits. Look also for score or scratch marks, which result from the wheel shifting on the shaft. If your engine is relatively new and you've never removed the flywheel, this type of cracking may be covered under warranty. It may also require replacement of the crankshaft because of damage to the shaft, keyway, or both.

Checking the magnets. Modern magnetic materials ensure that you will almost never find a weak magnet in your flywheel, unless your engine is very old. A quick and usually reliable check is to loosely hold a screwdriver blade about ½ inch from the magnets. The blade should snap over against the magnet strongly. This is an adequate indication that the magnets are good.

Checking for a loose main bearing. The rim of the flywheel passes close to the ignition coil armature and may show bright spots or other evidence of contacting the armature laminations. This could result from shifting or improper adjustment of the armature clearance—covered on page 136—but it might also indicate a very loose main bearing. Try pushing and forcing the flywheel and crankshaft towards the coil. If it's loose enough in the bearing to close up the normal coil-armature gap, the bearing needs repair.

This flywheel sheared a key and turned on the shaft, as can be seen from the score marks in the center. Never reuse such a wheel.

Flywheel-Brake Inspection

Whether the flywheel brake on your small gas engine uses brake pads or bands to bear against the flywheel, it also requires an ignition stop switch that automatically kills the ignition at the same time the bail is released and the brake applied.

Pad-type brakes. The type of brake with a pad of friction material that acts against the *bottom surface* of the flywheel is easiest to inspect and service. That's because the flywheel does not have to be removed and the mechanical and switch action can be easily observed.

This device—as well as others to be discussed— is actuated by relatively powerful springs. Before starting an inspection and repair, disconnect the battery completely to prevent shorts and keep your fingers from being caught in the mechanism. The spring used on the bottom-surface flywheel brake is

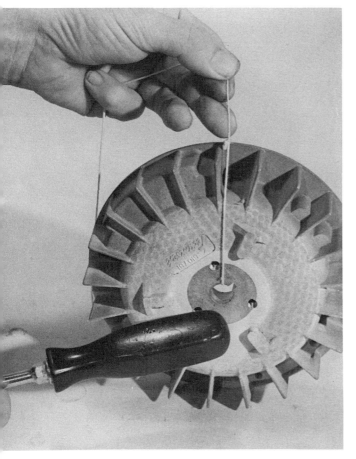

Check for flywheel cracks you can't see by suspending the wheel with a string and tapping it sharply. It should ring with a clear, bell-like tone.

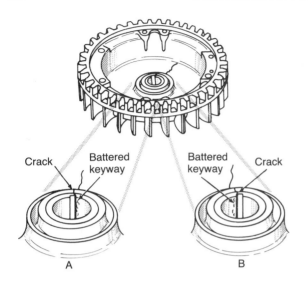

Crack Battered keyway Battered keyway Crack

A B

A flywheel cracked as shown in *A* is often the result of a mower blade striking an object. The crack shown in *B* may originate from a loose flywheel battering the keyway on each power stroke.

a torsion type; release the tension using a pair of hose-clamp pliers to free the short end of the spring from the projection that holds it. This will avoid stripping the starter-mount bolt threads in the engine block on electric-start engines.

Providing that the brake-pad material is worn no thinner than .60 inch at its thinnest part, and that the braking surface of the wheel has not been scored by worn or missing material, your primary concern is that the brake pad comes in firm contact with the flywheel rim and that the starter-interlock and ignition-stop switches are fully actuated. As with the throttle/governor control cable described in Chapter Eight, this is another case where a simple mislocation of the cable end can cause what appears to be serious trouble. The actual travel of the cable is controlled by the adjustment at the bail end. If this adjustment is slightly short, wear on the brake material may result in poor and ineffective braking. Or, the stop switch may not quite go full travel and you'll have no ignition. The same condition could prevent starter action.

Therefore, begin by making sure the safety-control cable slides freely and is not misadjusted, kinked, or rusted. Then test the make-and-break of the ignition stop and starter switches with a volt/

ohmmeter or continuity light. If either seems damaged or erratic, replace them both. Carefully note the ignition grounding clip and the routing of the wires, and be sure they don't contact the flywheel after they're reinstalled. If you suspect charging or internal ignition problems, follow the troubleshooting procedures laid out in Chapter 11.

The type of brake that bears against the flywheel *inner rim* requires removal of the flywheel for most service and inspection. Since the brake pad will normally be riding on the wheel rim, you must relieve the pressure before removing the flywheel.

1. Move the actuating lever to the *Off* position and lock it there by inserting a nail or piece of wire through it and the mating hole

Check the magnets by dangling a screwdriver near them. The screwdriver should jump over and stick firmly to the magnets.

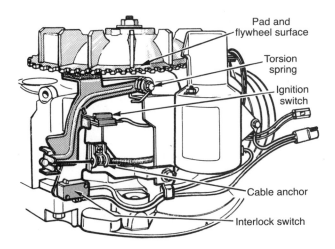

Pad and flywheel surface

Torsion spring

Ignition switch

Cable anchor

Interlock switch

Routinely inspect the operation of the blade-brake parts. Check the friction material on pad or band for wear or scoring. With electric start, in addition to the ignition switch check the interlock switch and cable anchor.

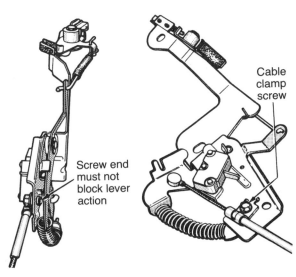

Cable clamp screw

Screw end must not block lever action

The cable-anchor clamp controls the movement of the brake and switch parts and must be secured and adjusted properly.

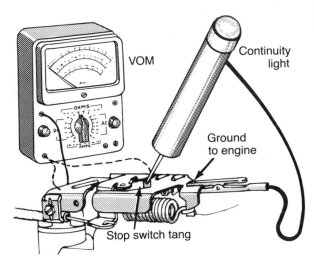

VOM

Continuity light

Ground to engine

Stop switch tang

Brake switches are easily checked for function with a VOM or simple, battery-powered light. If they are going full travel and don't have a positive on/off action, replace them.

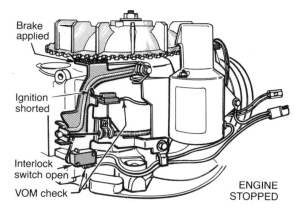

Brake applied

Ignition shorted

Interlock switch open

VOM check

ENGINE STOPPED

Remember that the ignition and electric-starter interlock switches work in opposite ways. The engine is in the *Run* mode when the ignition switch is *open* and the starter switch is *closed*.

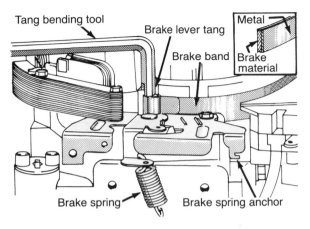

Tang bending tool
Brake lever tang
Brake band
Metal
Brake material
Brake spring
Brake spring anchor

On this engine, replacing the band-brake friction material requires bending the anchor pin tang to slip the brake free. Note that the brake spring must first be released.

at the top of the bracket. *Be sure the battery is disconnected before putting fingers or tools near the brake area.*

2. Inspect the brake pad. Again, if it is worn thinner than .60 inch at its thinnest part, the lever and pad must be replaced as a unit.

3. If the flywheel has been damaged by a worn pad, it must also be replaced following the procedure on page 112. Be sure to remove the nail or pin you used to hold the brake in the "Off" position and release the tension spring found on this type of brake from the bracket. The lever and pad are retained by a small E-clip on the pivot pin.

Band-type brakes. Although different in mechanical arrangement from pad brakes, the band brake functions in much the same way. It can be used with either manual or electric start. In either case you can get to it by removing the blower housing and the control-cover housing. With electric starters, the battery must be removed and some care is required when removing the switch housing.

Troubleshooting is essentially the same as with pad brakes. Check the spark by removing the spark plug or by connecting a plug tester in the high voltage terminal and grounding it to the engine. If no spark occurs with the safety control in the "Run"

position, disconnect the ground wire from the stop switch at the engine. If this produces a spark, look for improper actuation of the stop switch. The switch may be loose or the small tang that contacts the control lever may not be providing enough movement to operate the switch. With the plug wire disconnected and grounded, check the switch operation for positive on-off performance with a volt-ohmmeter. As with the pad brake, the operating control from the safety bail must fully actuate the switch and associated brake parts.

The brake-band material is bonded to the metal outer band. Inspect it for damage or thin spots. On band-type brakes, the manufacturer advises replacing the band if the brake material is thinner than .030 inch—or about the thickness of an ordinary book match. One end of the band is retained by an anchor pin, and the movable end slips over a tang on the actuating control lever. To remove the band:

1. Release the spring tension and bend the tang so the band may be slipped off.
2. Check to be sure that the anchor pin is secure and not bent or misaligned.
3. If the replacement band—with the tension spring back in place—allows the flywheel to rotate freely with the safety bail depressed, and if releasing the bail locks the flywheel enough to resist a torque-wrench reading of 45-inch pounds, the band adjustment should be satisfactory.

For a more accurate adjustment check, you'll need a special gauge-link tool. Adjustment is made by loosening the three screws that retain the control bracket to the engine. The bracket may be shifted right or left to alter the band setting. Be sure the spring is in place while doing this. If you have any doubts, do not try to operate the engine. Take the mower to a dealer for adjustment of this important safety control.

Ring-Gear Service

Most small gas engines do not have separate ring gears for starter engagement but, instead, have cast-in teeth which are part of the flywheel. With the frequent restarting on today's walk-behind mowers, it is possible for enough wear to occur so that the flywheel teeth and the starter pinion gear no longer mesh properly. In some cases only the starter pinion needs replacement; but if the flywheel teeth are severely worn or broken, flywheel replacement is the only option.

Some larger engines such as those used in tractors do have replaceable ring gears, which are sold together with the fasteners to secure them. These gears are not shrunk on as are automotive ring gears but are riveted and bonded to the wheel rim. To remove the gear from such a wheel:

1. Center-punch and drill out the rivets and then heat the wheel until the bonding softens and the old gear can be removed.
2. Scrape the old bonding from the wheel so the new gear will seat cleanly. You don't have to use bonding with the new gear if machine screw and locknut fasteners in the kit are used to secure it.

Flywheel-Fan Service

Nearly all small-gas-engine flywheels have integrally cast fans or vanes. Never operate an engine if you find one or more vanes cracked or broken. Doing so may cause serious imbalance and be dangerous. A few engines have separate plastic fan vanes that can be replaced.

1. Remove the old fan by tapping the rim.
2. Heat the lower portion of the new fan in boiling water until it can be slipped over the flywheel. *Be sure to align the locators with those on the wheel.*

    When the new fan cools, it will be locked in place on the wheel.

Clumsy attempts to keep this wheel from turning by holding a screwdriver in the blower blades broke this blade. Wheel is now unbalanced and unsafe to use.

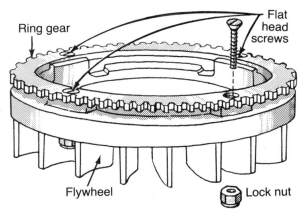

Worn or damaged ring gears can be replaced on some engines, but they must be selected to match the starter-pinion material. A Nylon pinion requires an aluminum ring gear while a metal pinion requires a steel ring gear.

10

Servicing Ignition Systems

For your small engine's ignition system to work properly, all that really matters is that it delivers a hot spark for starting and a reliable and properly timed spark while running. Though ignition systems have changed, often to solid-state, most of the basic operating principles remain unchanged. Here again, once you grasp these basics you can usually sort out the trouble and make a successful repair.

IGNITION-SPARK REQUIREMENTS

For the ignition spark to ignite the fuel that powers your small gas engine, it must have sufficient voltage—the term for electrical pressure—to jump the gap between the spark-plug electrodes. The spark must also provide enough amperage—roughly, the volume of electrical energy—to provide heat energy for ignition. An ignition system with leakage due to bad insulation, corrosion, or moisture might be compared to a water system that starts out with adequate pressure at the water meter, but loses part of the water along the line to leaks.

Finally, the spark must be located properly relative to the combustible mixture. Often an otherwise-good ignition system will skip and miss because carbon or other fouling material is providing a path for the spark up around the spark-plug core, rather than across the electrode gap. There is very little mixture available for combustion up inside the plug.

Magnetism. A few small engines have been built with battery ignition much like a car's. But a dead battery means a dead engine; thus the occasional use most small engines receive makes it much more attractive for their ignition systems to generate their own electrical energy. The key to that is the ability of a magnetic field to generate an electrical current when it moves past a wire conductor or the wire moves past it.

The all-important requirement to stimulating a current flow in a wire is *movement,* either of the magnet or the conductor. When the magnets cast or bonded into your engine's flywheel are at rest, they are putting out nothing. Grasp the spark-plug wire of your engine when it isn't running and you'll feel nothing. Turn the flywheel slowly and you'll probably still feel nothing, but turn it a little faster and you'll start to feel a small tickle of current. Spin it rapidly and you'll drop the wire in a hurry. This is sometimes called the *coming-in speed* of a magneto. Below a certain critical speed there is some power produced but it's not enough to overcome the internal resistance of the coil wire.

In practical terms that means that to start your engine you must spin the flywheel fast enough to produce a substantial current flow. It's one reason why a large engine may be hard to hand-start in the winter; the cold oil prevents you from pulling the flywheel around fast enough.

The high-voltage coil. A new plug in good condition might require only 5,000 to 8,000 volts to fire. A worn plug with a wide gap might require 15,000 to 20,000 volts. In either case, the relatively low voltage produced by the magnet movement will not begin to approach such high voltage. What's needed is a second electrical action to boost or transform the low voltage to a much higher level. That's the job of the high-voltage ignition coil.

The coil begins with a primary winding of fairly thick copper wire around an armature of soft iron or other easily magnetized and demagnetized material. As the flywheel turns, the magnets pass the coil and the magnetism is converted to electrical energy. The electricity flows through this primary winding and then through the breaker points, when they're closed, to electrical ground. This is the primary circuit, which produces low-voltage current. The primary winding consists of only a few hundred turns.

A secondary winding is then wound around the

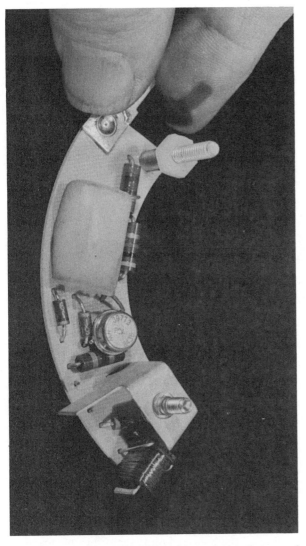

Shown at left is a solid-state ignition system before being encapsulated in epoxy. It performs the same job as the breaker points below—which have corroded and are starting to pit—yet has no moving parts and requires no regular service.

primary and consists of many thousands of turns of fine, hair-like wire. The magic here is that the voltage induced in the secondary winding is almost exactly in proportion to the difference in the number of turns of wire between the secondary and primary windings. Thus, for practical purposes, a few volts in the primary are transformed into several thousand in the secondary.

Two actions contribute to that transformation. First, the movement of the flywheel magnets past first one leg and then the second leg of the coil armature causes a rapid reversal in the magnetic lines of force. This raises the primary voltage, and when it is at its peak the breaker points open to cause

the current flow in the primary winding to suddenly collapse. The magnetic field, in collapsing, induces a much higher voltage in the many hair-like winds of the secondary winding. This happens because the magnetic lines of force are now cutting across *thousands* of turns of wire instead of merely hundreds. The more turns of wire the magnetic lines of force can cut across, the more voltage will increase.

In some small engines the coil and armature are mounted outside the flywheel, and the outer rim of the wheel runs very close to the legs of the armature. In other engines, the entire unit is mounted under the wheel, with the armature legs spaced closely to the magnets on the inner rim. The clearance between

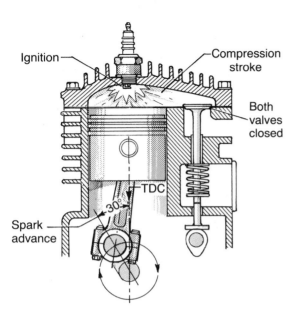

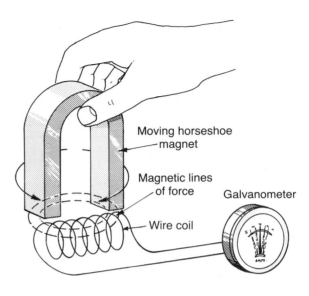

High voltage is needed for the spark to jump between the plug electrodes in the highly compressed air in the combustion chamber. Ignition timing varies but may occur as much as 30 degrees before TDC.

The familiar grade-school science demonstration of a horseshoe magnet that is waved across a wire and generates enough current to move a meter needle applies to your engine.

the armature legs and the magnets is called the armature air gap, which in many cases is adjustable. As I'll show later, the gap must be adjusted wide enough so there is no possibility of the flywheel striking the armature legs, yet close enough to make maximum use of the magnetic action.

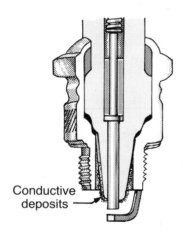

If conductive deposits build up on the core nose and inside the plug, the electrical energy may bleed off without a spark.

BREAKER-POINT SYSTEMS

Always remember that it is the *opening*, not the closing, of the breaker points that times the firing of the spark plug. Many who thought of the breaker points as a switch that directed current to the plug have tried, unsuccessfully, to time their engines on point closing.

The points must be closed for a significant length of time before opening, however. This is because there must be time enough for the magnetic field and primary current to build up to an adequate level. Like any wire, the primary winding has two ends. One end is grounded to the armature or engine structure. The second end is connected to the moveable breaker point. When that point is closed and in

Above and below: coil and magnets may be inside or outside the flywheel, but both must be closely spaced.

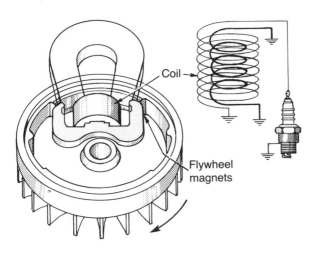

contact with the grounded fixed point the circuit is complete between the two ends of the primary coil via the engine structure.

Although the cam mechanism that opens the points is timed for the proper moment of point opening, this timing will be correct only when the breaker point clearance is correct. If the points are set too close they will open early, thus advancing the spark

timing relative to piston position. If they are set too wide the timing will be late. On the majority of small engines, the breaker points are located under the flywheel. Some engines, however, have them located externally in a small housing on the side of the engine. The condenser is usually mounted near that housing.

The condenser. Although it is technically more accurate to refer to a condenser as a *capacitor*, the former term is still more commonly used when speaking of that small, usually tubular-shaped component of the ignition system. When the breaker points suddenly open and break the circuit through the primary coil, there is a large amount of electrical energy that must be dissipated quickly. Without a condenser, much of this energy would find its way across the breaker points in the form of arcing. That would cause rapid point burning and, because it would slow the collapse of the primary field, it would also hamper the development of high voltage in the secondary. The condenser is a handy recipient of this energy and is connected between the moveable breaker point and ground.

The condenser normally consists simply of two rolls of metal foil separated by insulating paper. One foil surface has the ability to accept a charge of electrons, while the other foil is oppositely charged. When the breaker points close again the condenser is again grounded out. Remember that a magneto system, or a solid-state system energized by magnets, is exactly the opposite of a battery system. Any switch in a small engine system must be *open* for ignition to occur. Anything that establishes a circuit between the primary circuit and the engine or machine structure—from an ignition switch to a bare wire, including moisture—will prevent ignition. Thus, if a condenser, which has one side grounded to the engine frame, has an internal short that allows the passage of electrical energy between the two foil surfaces, it's equivalent to closing a switch to stop the engine or prevent starting. If you find a second wire connected to the condenser terminal it will almost always be a wire running to the stop switch.

The capacity of the condenser in your engine to absorb and store electrical energy is expressed in microfarads (μf), and this is usually stamped on the outer shell. Burned breaker points are usually a certain indication of a defective condenser.

The secondary system. The only part of the secondary or high-voltage circuit you're likely to be well acquainted with is the heavily insulated lead wire from the coil to the spark plug. In early engines the plug wire was often soldered or just twisted onto the outlet terminal on the side of the coil. Today, the heavy wire will probably be cemented or molded in place and cannot be replaced readily. There is no real service for this wire with the exception of inspection for insulation chafing or cutting, which may happen if you're not careful when you reinstall the blower housing. Usually there is a dimple or channel into which the wire must go before the housing is tightened.

The plug end of the high-voltage wire may have a variety of terminal fittings ranging from a simple fork-shaped clip to a rubber boot shield with a spring grip up inside. Abuse of the rubber boot by tugging or pulling on the wire often causes poor contact. That means the spark energy either has that much more resistance to overcome, or it can't make it to the plug at all. Before invading your ignition system because you get no spark at the plug, be sure that you have checked the high-voltage wire and connection at the plug.

Unit-Type Magnetos

Although most small engines today use some form of flywheel magneto ignition, either breaker-point or solid-state, you may encounter a conventional unit-type magneto on heavy-duty farm or industrial engines or on older engines. You'll immediately recognize a unit-type magneto since it mounted externally on the engine and is driven by a gear or coupling engagement. On some old engines it may be mounted on a shelf-like base, but most are mounted by two capscrews through a flange on the front that matches the engine drive flange.

The same principles of magnets and movement that apply to flywheel magnetos also hold true for unit magnetos. The very old ones with the familiar horseshoe magnets had a wire-wound rotor that turned inside the magnetic field established by the magnets. More recent types use a magnetized rotor turning inside an armature and coil structure similar to the flywheel types. A cam on the back of the rotor shaft opens and closes the points and produces a secondary current in conjunction with the condenser, as it does with more traditional breaker-point flywheel ignition.

Cutaway of a high-voltage coil reveals the coarse primary winding around the center and the extremely fine, hair-like secondary winding surrounding it.

Impulse couplings. Unit magnetos must be turned at fairly high speed to produce a spark. Since they are usually driven off the camshaft gear, which turns at only half crankshaft speed, the cranking speed is seldom fast enough to produce a good starting spark. Not only are most engines using unit magnetos started manually, they are also fairly large as a rule. Add to that the normal advanced spark timing before top center, and those engines are capable of kicking back dangerously.

A device called an impulse coupling is commonly used both to develop the necessary rotor speed for starting and to retard the spark to just past top-center until the engine is running. The impulse coupling is housed in a small, cannister-like shell at the point where the magneto is coupled to the engine. You'll know it's there because if you crank such an engine over slowly you'll hear a snapping sound as the piston passes top-center on the start of the power stroke. This is caused by a husky torsion spring inside the coupling. At cranking speeds a tang engages a stop and, although the magneto drive turns, the magnet rotor is held stationary.

As the drive rotates further, a trigger releases the now tightly wound spring and the spring snaps the rotor around quickly, thus providing the rapid movement of the magnets needed to generate pri-

How The Coil Works

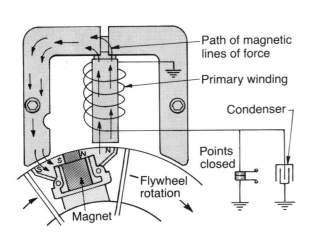

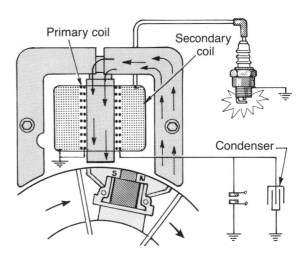

1. The high voltage needed to cross the spark-plug gap begins when the magnetic lines of force from the flywheel magnets flow through the center leg and one outer leg of the coil core. Primary current starts to build. Points are closed, and each end of the primary is grounded, providing a complete circuit for primary current to flow through coil.

2. As the magnet poles move past the second outer leg of the core, the magnetic flux reverses polarity and primary current builds still higher. The breaker points now open and the primary current no longer has a continuous circuit and collapses—cutting the secondary windings and inducing high voltage that discharges through the spark plug.

mary current. Moreover, during the time delay before the coupling is snapped, the piston has reached or gone a trifle past top center, thereby reducing the chance of kickback. The action of the impulse drops out by centrifugal action on an inner weight as soon as the engine begins to run at slightly more than cranking speed.

SERVICING BREAKER-POINT SYSTEMS

You'll find the breaker points under the flywheel on most engines, though some engines have separate housings on the side of the crankcase.

If the points on your engine are located beneath the flywheel, begin by following the flywheel-removal procedure on page 112. Once the wheel is off, examine the breaker points to determine whether they need replacement and, if so, what caused them to deteriorate. Was it a poor con-

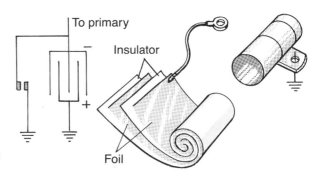

The symbol for a condenser is shown at left and represents two plates separated by an insulator. Placed across the breaker points, the condenser absorbs the current surge when the points separate and prevents arcing and point burning.

The inner parts of a unit magneto look and work the same as in a flywheel/breaker system. Note the points, condenser, and coil.

The drive-end of a unit magneto will usually have an impulse coupling. As you turn the engine to start, the tang holds back armature rotation until the piston is slightly past top center. Then a spring action snaps the armature around to produce a hot spark.

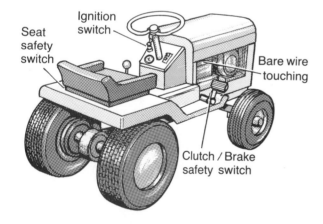

Most tractors and rider mowers have safety interlock switches under the seat and in the brake/clutch linkage or shift systems that can ground out ignition. Also check the ignition switch and the wiring that runs through and around the frame.

denser—as might be indicated by severe burning—or has excess oil, moisture, or corrosion been a factor? Perhaps the breaker-point *rubbing block,* which runs on the cam, has worn and the clearance has merely closed up.

Breaker-Point Cleaning and Inspection

Often you'll find that the points appear to be in good shape but you aren't getting a spark. One common reason is long storage, which may have caused an oxide buildup that prevents good electrical contact.

1. Slip a strip of clean note paper between the contacts and turn the shaft so the points close on the paper.
2. Slowly withdraw the paper. If it comes out with a dark streak on it, you've removed some of the deposit.
3. Now run a fine, thin ignition file between

the points—take no more than two or three strokes.

Now try slipping the wheel back on and snugging down the nut finger-tight. Spin the flywheel with the lead wire attached to the spark plug and the plug removed from its hole. If you see a spark at the plug, you've probably solved the problem.

Oil-fouled points. The next most likely condition you may find is oil contamination. Although most engine parts need plenty of lubrication, electrical contacts should be clean and dry. On most small engines the breaker points are located very close to the crankshaft oil seal. Also, on Briggs & Stratton engines, the points are actuated by a small pushrod or plunger that extends through a hole in the crankshaft main bearing and rides on the crankshaft. A flat on the crankshaft serves as a cam. If even small amounts of oil leak around the crankshaft seal or around the little plunger, the oil can accumulate in the breaker area and cause ignition problems.

The crankshaft seal, or the plunger and the bore it runs in, may be worn enough to allow oil to es-

cape. But don't overlook the possibility that the crankcase breather system may be clogged, gummed, or leaking. Excess crankcase pressure may be forcing oil into the breaker area. Another possibility, especially with an old engine, is that the rings, piston, and cylinder bore may be worn so badly that the blow-by into the crankcase exceeds the breather capacity. You may now, or later, want to run a compression or vacuum check to see if this is true.

If the problem is a worn plunger bore, the repair involves disassembling the engine and removing the crankshaft, reaming out a bore for a service bushing and pressing in the bushing, then reaming the bushing to size. Not only is this a dealer-shop job, the need for special tools as well as the parts and labor cost involved make occasional cleaning of the breaker area a more attractive alternative.

To clean the breaker area use a small brush and lacquer thinner or other solvent to remove the oil so that it may be wiped and blown away. *Do not use carburetor cleaner or any vigorous solvent on the crankshaft seal, which is in the same area.* If the leakage is coming from the crankshaft seal, you can

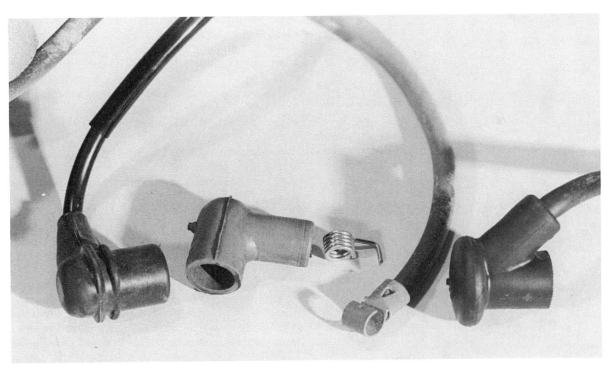

Most spark-plug boots contain a spring-like connector that pierces the wire insulation for contact. Often, these parts are difficult to reassemble if pulled apart. Use a boot tool to prevent this.

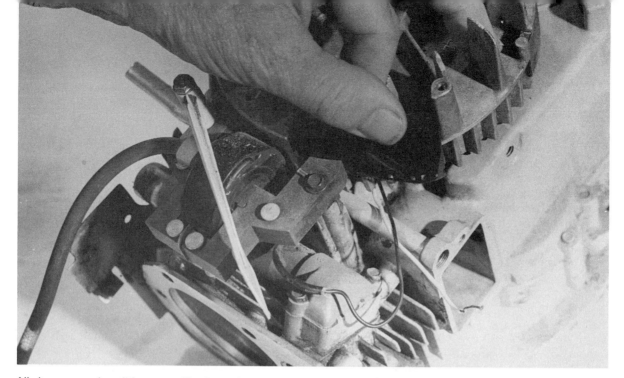

All shop manuals call for a specific air gap between the magnets and coil laminations of this type of assembly. You can usually use a layer or two of photo film as a gap gauge.

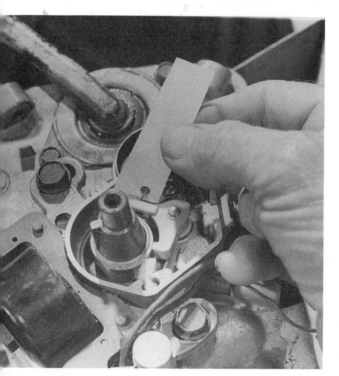

Try closing the breaker points on a piece of bond paper and then withdrawing the paper. If you find oil like this, that's probably why you're not getting a good spark.

remove and replace the seal if the main bearing extends out through the breaker housing. If the seal is down under the stator assembly, you will probably find that removing the stator for access to the seal will require retiming the engine by shifting the stator on reassembly. Again, unless you are prepared to go this far—or the engine is vital to your livelihood and must be run every day for extended periods—it may be better to simply clean things up when oil accumulates.

Reduced clearance. The next thing to look for before you replace the points is whether the point clearance has closed up until the points barely separate. When this happens it usually means that the breaker arm has worn badly at the point where it runs on the cam. This part of the breaker arm, called the rubbing block, may be a small piece of plastic material attached to a metal arm, or simply a projection molded onto a plastic arm.

Excessive wear may be caused by fine dust or dirt, or a cam that has rusted or become rough. In most breakers, a small piece of felt rides on and wipes the cam profile to keep it clean. A factory lubricant has been saturated into the felt and should provide the slight lubrication needed. In most cases there is no need to re-oil this wiper; doing so will introduce unwanted oil into the breaker area.

In some engines, a second cause of limited point

Briggs & Stratton makes use of this fiber push rod to actuate the breaker points.

opening can be a worn plunger. Replace the plunger by drawing it out and slipping in a new one. Usually the small groove on the plunger goes towards the outside. Reversing it will cause oil leaks.

Breaker-Point Replacement

The breaker-point assembly you'll usually see is held down by one or two screws and removed as follows:

1. If the points have a flexible wire leading to a terminal block where the wire is joined by the coil primary wire, loosen and remove the two wires and carefully inspect the block and screw for oil fouling, cracking, or other conditions that might provide a path for current leakage.

2. If oil is found, remove the terminal block and wash and dry the parts and area surrounding it.

3. With the wire from the points disconnected, removing a second screw allows the old breaker-point assembly to be lifted free so the new one can be installed.

Adjusting point clearance. With the new points in place but the retaining screws not quite snugged down, your next step is to adjust the point clearance. Typically, clearance will be about .018–.020 inch. If the gap is too great, the points will not remain closed long enough for primary current to build up fully. If the gap is too small, the points will open at the wrong time relative to the magnetic flux and the primary current. Ignition timing will be somewhat off and power will be lost.

Although details vary, the most common adjust-

It's good practice to wash the breaker-point area clean with solvent and blow it dry before installing new points. Otherwise, old grease and other debris may find its way onto the point contact surfaces.

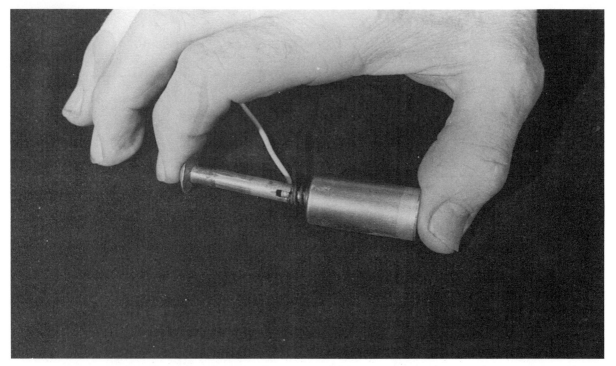

This condenser incorporates a fixed breaker point as one connection and has a spring to clamp the primary wires in a small hole. The spring-compressing tool at left—made by drilling out and slotting a carriage bolt—helps in getting the wires through the hole. Many such tools are easily made in the home shop.

To remove the old points, begin by removing the small nuts and washer that retain the two wires connected to the terminal block. A ¼-inch socket set comes in handy for such close work.

ment procedure requires careful movement of the fixed point with the retaining screws slightly loose.

1. Begin by turning the crankshaft until the breaker-point rubbing block is on the high point of the cam, positioned for maximum separation of the points.
2. Measure point clearance by slipping a wire feeler gauge between the fixed and movable points.
3. Shift the fixed point by gently prying it with the tip of a screwdriver until the feeler gauge can be inserted or withdrawn with just a slight drag. *Remember—the spring pressure on the moveable point is light, making it easy to inadvertently move the points apart with the gauge and set them too close.*

On breaker-point assemblies with one point integral with the condenser, adjustment requires sliding the entire condenser forward or backward in the mount. Once the adjustment is made, snug down the retaining screws and recheck the clearance. Often the tightening action will cause the setting to change.

Although the setting on most small engines is not extremely critical in terms of performance, a much more accurate setting can be made by fixing the sensing contact of a dial indicator against the moveable point and rotating the crankshaft. This will give a true reading of the actual clearance and may also reveal such problems as crankshaft wobble caused by loose bearings.

Once the new points are installed and adjusted,

Gently lift out the terminal block and moveable breaker arm and spring attached to it.

clean them of any oil or fingerprints by drawing a piece of clean paper between them. Inspect the points to be sure that no lint or fragment of paper remains, then reinstall the flywheel. Remember—the original flywheel key should be used only if it is snug-fitting and shows no signs of shearing or battering. Again, *never replace a soft-metal key with a steel key.*

Sealing out moisture. If the points on your engine are covered by a snug-fitting box of thin metal, note that the exit of the coil wire will probably show a deposit of adhesive sealant that keeps out dirt or moisture. Use silicone sealer to reseal this point and any others that appear less than well-sealed. If the manual calls for ignition retiming, or you have removed the breaker stator-plate, do not seal down the cover until after timing.

Servicing External Breakers

Replacing points on engines with the breaker mounted in a housing on the side of the engine is easier than replacing those requiring flywheel removal. This location does, however, present an even greater opportunity for oil to leak from the crankcase and onto the points.

After removing the housing or cover, inspect the breaker area for signs of significant oil leakage. In some engines a seal will be found around the actuating plunger, which comes through the side of the crankcase. For some other engines without such a seal, it is possible to add one.

Other solutions to excessive leakage require removal and replacement of the plunger and plunger bushing in the crankcase wall. One type of plunger

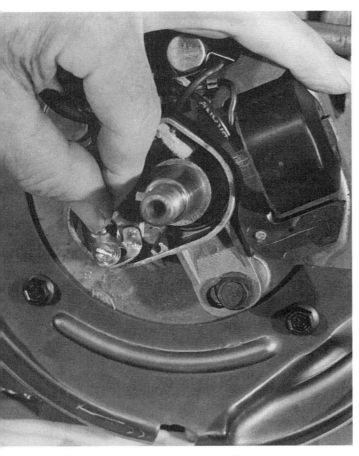

Remove the screw that secures the fixed breaker point and then lift out the fixed point.

bushing has short threads projecting from the crankcase; by threading a nut against a series of washers between the nut and the case, the pressed-in bushing can be progressively pulled out. A second type of bushing lacks threads, which means the plunger must be broken off flush with the bushing and the bushing tapped for puller threads that can then be used with a short stack of washers to withdraw it. In both those procedures, take care not to push the plunger or metal cuttings into the crankcase.

Replace the plunger and bushing assemblies by carefully tapping them into the crankcase to the proper depth. Use a hollow-center driving tool such as a scrap piston pin or the old bushing to drive them.

Sometimes removal of an external breaker assembly requires removing screws that penetrate the crankcase wall. On reassembly, use a sealant on the screws to avoid an oil leak. Gap adjustment is similar to that for breakers beneath the flywheel. Again, if retiming is necessary, do it *before* sealing the box.

Testing A Condenser

After you remove a pair of old points, examine them for cratering and transfer of metal from one point to the other. Those are good indications of a defective or incorrect condenser. It's easy to make a preliminary check of condenser condition with a volt/ohmmeter.

As with spark plugs, setting the gap on breaker points should be done with a *wire* feeler gauge, not a flat one.

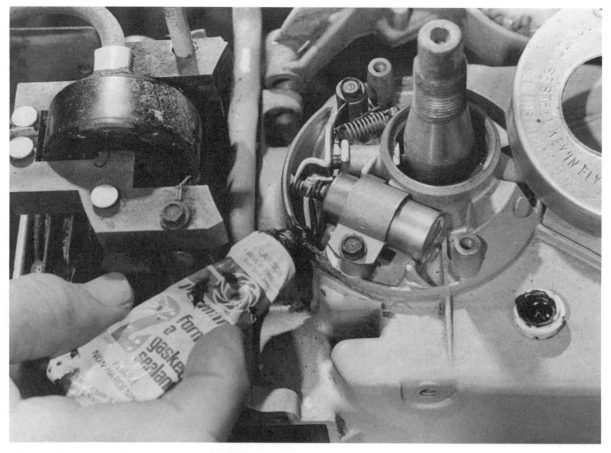

The breaker housing on some engines is sealed at the wire exit. Clean away the old sealant and reseal as shown. This prevents moisture, dust, and other debris from getting inside.

1. Set the volt/ohmmeter (VOM) on the ohms scale for a continuity check.
2. Touch one probe to the wire terminal and the other to the outer housing of the condenser. The condenser should be disconnected from the breaker points.
3. The VOM needle should quickly kick up to show charging of the condenser and then slowly drift back towards zero current passage. Reversing the probes should produce the same action.

This test usually indicates a healthy condenser, though heat or vibration from the engine might be causing it to short out while running.

If the VOM needle does not move, the condenser is defective. Or, if the VOM needle moves over to show current passage and holds there, the condenser is shorted internally. The action of the needle as it kicks over initially represents one plate of the condenser taking a charge of electrons. When the plate becomes charged, there can be no more electron flow and the needle drops back.

An interesting sidelight presented in some manuals is that a buildup of metal on the moveable point may indicate a condenser with too little capacity, while a buildup on the fixed point may indicate excess capacity. Such diagnostics may impress your friends and neighbors; but it's still best to stick with the specified condenser.

Checking The Coil

As with the primary coil winding, the secondary winding also has two ends. One is the plug lead and the other may be buried and connected to ground inside the coil or may be brought out and grounded together with the primary lead outside the coil body.

On the theory that if a wire isn't broken, electrical energy applied at one end should come out the other end, you can make a simple continuity test on the secondary winding with a VOM.

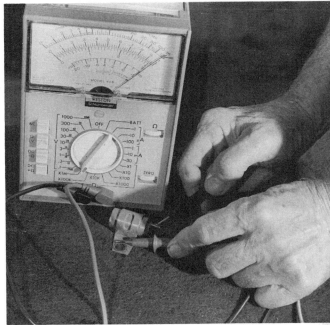

A quick and usually reliable condenser test can be made with a VOM. With one probe on the shell and the other touching the connector lead, expect the needle to kick over and then fall back slowly. If you can reverse the probes and get the same result, the condenser is probably okay.

1. Begin by separating the coil leads from the condenser lead.
2. Touch one probe to the metal spark-plug connector and the other to any bare metal on the engine. The needle should move substantially to indicate continuity.
3. The actual ohmic resistance of the secondary winding should be substantially higher than that of the primary.

The idea of this electrical test is to isolate a single circuit at a time without other components connected, which might introduce variables. The test will usually reveal a broken secondary wire or bad plug lead, but it may not if the break only appears when the coil is warm. The test will *not* clearly show failure of the coil insulation and shorted internal windings. If the appearance of coil trouble such as intermittent missing, skipping, and low power persists, the coil must be removed and taken to a shop

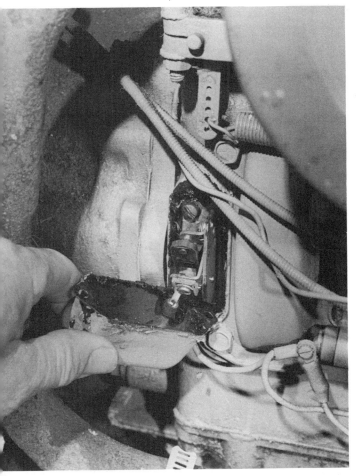

Breaker points on this engine are located down on the side of the crankcase, not under the flywheel. Point operation is via a push rod extending through the side of the crankcase.

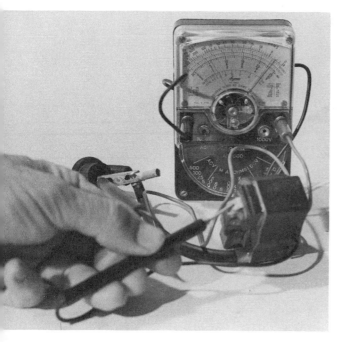

A VOM continuity check between the coil-ground and spark-plug connections will tell you if you have a solid connection through the coil and to the plug. Take a specific ohms resistance reading and check it against a new coil for more insight.

Some engines have slotted mounting holes for adjusting the ignition-coil armature. Refer to a shop manual before loosening these bolts.

with a coil tester to load the coil electrically and test for actual performance.

Armature Air-Gap Adjustment

Many small engines make no provision for adjusting the timing of the ignition spark. If careful examination of the coil armature shows no slotted mounting holes, and the parts supporting the breaker plate or stator show no means of shifting them, you can probably replace points or otherwise service the engine without concern. But if you see slotted mounting holes for the stator—and by slotted I mean the holes are elongated a little, rather than perfectly round—think twice before loosening any of those parts.

If the engine has been running well, no one has tampered with it, and the bearings aren't excessively worn, the air gap between the coil armature and the flywheel should be correct. The air gap has a definite relationship to magneto and electronic-ignition performance and is the starting point for timing. Measure it with a strip of brass shim stock or plastic of the proper thickness slipped between the flywheel magnets and the armature pole shoes. Trying to use steel feeler gauges is very difficult because of the magnet action. Two layers of 35mm black-and-white photo film closely approximate the .012-inch air gap frequently called for, though the specified gap does vary. Check the manual for your particular engine.

Ignition Timing—Internal Breakers

The ignition system on your small gas engine is one of many types, each with its own specific procedure. Basically, however, with mechanical breaker points you are aiming to time the magneto itself for the relative position of the magnets and pole shoes, and to time the separation of the points relative to piston position. Before going any further removing or adjusting the armature or stator plate, take a sharp steel scribe and mark the present position of all parts. Thus, if you encounter problems, you can always come back to where you started.

Although hard to see, a clear plastic gauge is being used here to adjust clearance between the coil arma-ture and the magnets. When the clearance is estab-lished, tighten the armature retaining bolts securely.

You'll find that most engine manuals call for reference to various marking locations. For example, on some engines there are marks on the flywheel and armature bracket that must be aligned at the same time the points are separating. In other engines, no marks are provided but at the moment of point separation the magnet insert in the flywheel must align with the edge of the armature pole shoe. Since the breaker points are under the flywheel, you'll need a VOM, or a flashlight battery and a bulb, to tell you when the points are separating under the wheel. Timing consists of loosening the armature slightly and shifting it on the bolts. Remember, if the engine was previously running satisfactorily and you have not removed the armature or stator, adjusting the point gap should be all that's needed.

In Tecumseh engines, timing is adjusted relative to the piston position as measured by a dial indicator through the spark-plug opening. The specification for the piston position is given in thousandths of an inch before top center. If you have a dial indicator

and can rig an adapter—perhaps with an old spark-plug body so the sensing part of the indicator can bear against the piston top—this is the way to do it. The task is more difficult on the typical small four-stroke engine because the plug opening is offset rather than directly above the piston.

If you have the cylinder head off for some reason, you can use a dial indicator to locate the timing point of the piston travel and then mark the flywheel and a handy adjacent point of the crankcase so in the future you can simply time to the marks. In general, however, leave the breaker stator alone and the timing will probably work out fine unless someone has moved it from the factory setting.

Timing External Breakers

On most engines with external breakers, adjusting the point gap is all that's needed to time the ignition. One maker, Kohler, recommends using a timing

If you must remove a coil assembly with slotted holes, scribe a mark on each side so you can relocate the armature in exactly the same position. If in doubt, check the manufacturer's specs.

light to view the marks on the flywheel while the engine is running. The procedure is as follows:

1. Remove the small cover plug from the back of the blower housing or engine end plate so the flywheel is visible.
2. With the spark-plug lead disconnected, turn the crankshaft in the direction of normal rotation while watching through the opening—use a flashlight if necessary.
3. When the "S" or "SP" mark—for *spark*— appears, dab it with a little white paint or chalk for visibility.
4. Connect your timing light in the usual manner using an automotive battery if the light requires one. Also reconnect the spark-plug lead in the proper manner for your light.
5. Now start and run the engine at normal

Tecumseh gives timing-advance specs in terms of piston travel in thousandths of an inch BTC, as measured by a dial indicator.

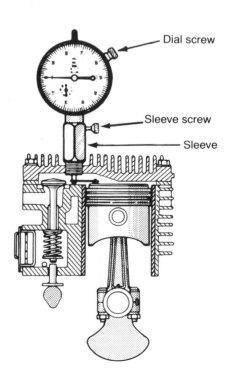

Dial screw

Sleeve screw

Sleeve

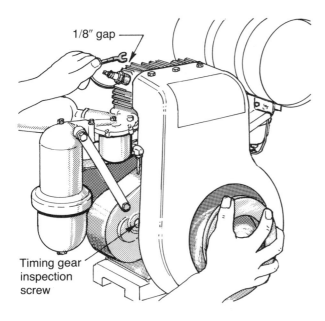

This engine also has ignition timing adjusted according to piston height. With the head off, measure piston position with a feeler and then scribe your own timing marks on the flywheel and crankcase.

1/8" gap

Timing gear inspection screw

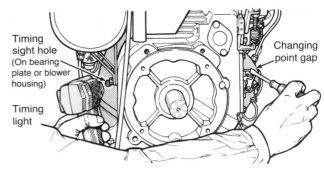

Timing sight hole (On bearing plate or blower housing)

Timing light

Changing point gap

Here, the flywheel timing mark is visible in the bearing-plate opening and is used with an automotive timing light as shown.

medium operating speed. The "SP" mark should be centered in the hole. If not, minute timing corrections can be made by slightly altering the point clearance.

If you don't have a timing light, the following method will also work:

1. With the flywheel cover plug and spark-plug lead wire removed, again, turn the crankshaft in the direction of normal rotation while watching through the opening. After the "S" or "SP" mark appears, continue turning and you'll see a "T" for *top center.* The distance between the two marks is the ignition spark advance for that engine.

2. Turn the engine another revolution in normal rotation until the "SP" mark has passed the opening slightly. If the breaker-point clearance is correct, the points should now be slightly open.

3. Slip a piece of thin cellophane between the points and turn the engine backwards at least a quarter turn until the points close on the cellophane.

4. Now turn the engine in the direction of normal rotation very slowly and maintain

To check for spark on a unit-type magneto, hold the high-voltage wire near the cylinder head and rotate the crankshaft in the normal direction by hand. You should hear a pronounced snap, and a spark should jump when the impulse coupling actuates.

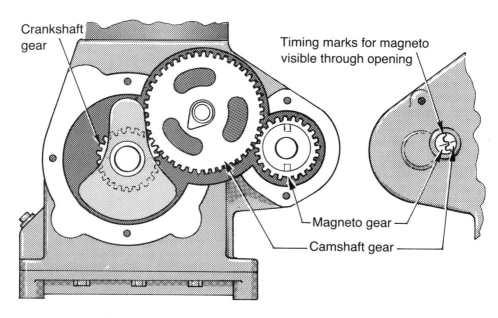

Installing a unit magneto requires matching the marked gear teeth on the cam gear and magneto drive gear.

light finger tension on the cellophane. Just as the "SP" mark centers in the hole, you should feel the cellophane release as the points open.

Again, if the points open too soon or too late, change the point clearance slightly to correct the timing. Remember to always back up and then turn in the direction of normal rotation when making the cellophane check. This will eliminate any errors that might be caused by gear backlash and running clearances. Note also that most external breakers are actuated by the camshaft, which turns at half engine speed. If the "SP" mark is aligned in the hole but there is no breaker-point opening action, you are at the end of the exhaust stroke and the crankshaft must be turned one more revolution to bring the piston up on the compression stroke and open the points.

Timing Impulse Magnetos

You are unlikely to encounter a unit-type magneto with an impulse coupling unless you're working with a heavy-duty small engine or a rather old one. Timing is not difficult, but it is different. Before

starting, set the breaker points to proper clearance.

Unit magnetos are driven off the camshaft gears at half engine speed. That means such engines will fire only every other revolution of the crankshaft and must be timed on the compression stroke. If you accidentally time a unit-magneto engine to fire on the exhaust stroke, the engine will not run; therefore, place your thumb over the spark-plug hole so you can turn the shaft and feel for compression buildup as the piston approaches top center on the compression stroke.

Also, the flywheel—or perhaps a pulley on the front of the crankshaft—will probably have a timing mark and probably a TDC mark. Look for such marks as "S," "SP," "IGN," "FIRE," or the like. On some engines such as Wisconsin, removal of a plug on the crankcase face opposite the magneto allows viewing a timing mark on the cam gear that must be meshed with a mark on the magneto gear. In any case:

1. Rotate the crankshaft slowly in the normal direction of rotation to align the mark with a pointer, if there is one, or with the center of the observation hole. Again, you may have to remove a cover or plug or the blower housing to find these timing marks.

2. The simplest way to release the impulse coupling to proceed with timing is to merely turn the coupling backwards until the cam and points are in the just-opening position. When turned in reverse, the coupling will not engage and you avoid the potential for a shock or skinned knuckles.

3. Before turning the coupling backwards slightly to close the points, slip a narrow strip of cellophane between them. Now gently rotate the coupling just past where the points close on the cellophane until the cellophane is released.

By experimenting a bit, you'll determine exactly where the points separate. This is the point at which you want to hold the coupling from further rotation and engage the drive gear or coupling.

A straight-tongue coupling seldom presents a problem. Unless the engine has been disassembled and the drive coupling improperly installed, canting the magneto slightly one way or the other should permit engagement. Helical-cut drive gears may present more of a challenge. When you present the teeth of the magneto gear to the internal gear on the engine, the helix will cause the rotor to turn and your nicely adjusted point opening will be upset. The trick is to observe just how much the rotor turned and start over again, this time with the rotor turned back a tooth or two so that when the magneto seats, the points are near or just opening. It may take several tries. If the cam gear and magneto gear are marked, the job will be easier.

You still have to make a final timing adjustment after snugging down but not tightening the magneto mounting bolts. Often, either the engine or the magneto will have a slotted bolt hole to permit some rotation of the entire magneto. Again, place the cellophane between the points and gently bump the magneto body with the heel of your hand in whichever direction is required to place the points in the just-opening mode.

Summed up, the target is to have the engine on the timing mark, on the compression stroke, and the points just separating without the impulse coupling being involved.

Timing Two-Stroke Engines

All of the preceding assumes you're working with a four-stroke engine. In the case of two-stroke engines, timing is particularly critical. If it is necessary to disturb the breaker-point assembly—if your engine has one—be very careful to mark the location of the parts before removal, and to replace them exactly the same way. This is a situation where the sheer variation of systems and procedures mandates a manual for your particular engine make and model.

SOLID-STATE SYSTEMS

There are many detail differences in the electronic circuitry and components used by different manufacturers to accomplish essentially the same results as were described for breaker-point ignition. Nevertheless, most of what we covered still pertains. The magnets remain in the flywheel and a primary current is still generated to power the system. The coil still consists of a primary and secondary winding, and the switch must be open for ignition to occur. The significant difference is the use of solid-state devices rather than mechanical points to interrupt the primary current flow.

Solid-State Advantages

In addition to eliminating the breaker points, solid-state systems have three major advantages: faster rise time, lower turn-on speed, and higher voltage output at both cranking and running speeds.

Rise time is simply the time required for the ignition system to build up the voltage necessary for the spark to jump the plug gap. Typically, a conventional breaker-point system may require 15 to 20 microseconds of rise time, while a solid-state system may require only 10 microseconds. If, during that time, part of the available energy in a breaker-point system leaks away via carbon or other conductive deposits on the plug, the needed voltage may never be reached and the plug will misfire. With electronic ignition, the faster rise time will fire a partially fouled or worn plug before too much energy has a chance to leak away.

Earlier I mentioned the coming-in speed of a magneto and pointed out that sometimes it's difficult to crank an engine fast enough to produce the needed ignition voltage for a starting spark. In an electronic system this is called *turn-on speed,* and may be as low as 120 rpm compared to 300 rpm for breaker-

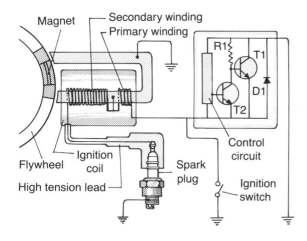

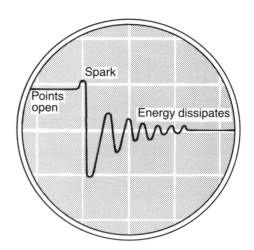

A typical transistorized ignition system. In operation, transistor T1 controls primary current until T2 senses proper current buildup. T2 then becomes conductive and primary current is bypassed.

A typical ignition spark on an oscilloscope. As the system is triggered, energy rises until the spark jumps the gap. After jumping the gap, the coil energy gradually disipates in a series of oscillations.

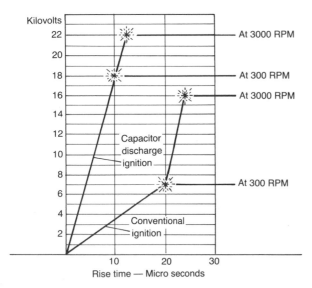

The time it takes for a breaker-point magneto system to develop proper voltage—its *rise time*—is longer than with electronic ignition.

point systems. In addition to providing energy for starting with less cranking effort, a good solid-state system can often produce twice the voltage at both low and high engine speeds. Thus, with more use-able voltage for starting at a much lower cranking speed and higher voltage at all speeds, your small engine is easier to start and more reliable overall.

Solid-State Components

The name "transistorized" is one of the terms that became more or less generic with the confusing array of solid-state devices. In truth, the system used in your engine may contain one or several transistors, thyristors, silicon-controlled rectifiers (SCRs), and Zener diodes, along with several conventional diodes. Indeed, all of these functions may be contained in a single microchip about the size of a postage stamp with which you can replace the points and condenser in your breaker-point ignition.

All of those devices control current in specific ways. Ordinary diodes allow current to flow in one direction but not the other. Place such a diode in an alternating current circuit and pulsating direct current comes out; only half of the current can pass the diode, and that is the half going the desired direction. We'll discuss that a bit more in Chapter 11 under *Alternators*. Other devices, such as thyristors,

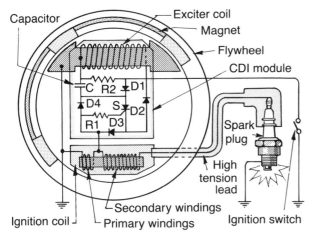

Capacitor discharge ignition. Here, energy built up in the exciter coil by the magnet action is rectified and used to charge capacitor *C*. When a trigger signal from the exciter coil turns thyristor *S* to a conductive state, the capacitor discharges through the primary and is transformed to high voltage.

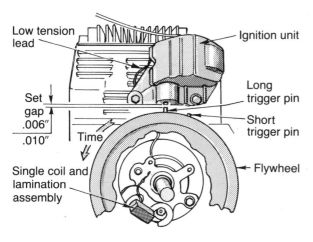

About the only service possible on this CD system is checking the trigger-pin clearance and readjusting the ignition unit as needed.

SCRs, and Zener diodes, perform other jobs, such as blocking current until a certain voltage is reached. A thyristor, for example, is combined with a transistor in the primary circuit of the coil. The magnetic power buildup and primary current flow are established until at a certain point the blocking transistor

or thyristor instantly turns conductive and, in a sense, dumps the primary circuit to ground. The collapse of the primary causes an induced high voltage in the secondary exactly as if a set of breaker points had opened and the primary had discharged into ground and a condenser.

Capacitor-Discharge Ignition

Earlier I said that the term *capacitor* was the technical reference to the ignition condenser. In talking about capacitor-discharge ignition, I'll refer to the same device as a capacitor because in this application it does, indeed, have the capacity to hold electrical energy in storage.

As with transistorized ignition, CD ignition uses solid-state devices to develop and time the high-energy spark discharge. CD does not use the collapse of the primary field to energize the secondary, however. Instead, at a properly timed instant, a high-energy charge is discharged into the primary and this energizes the secondary. The same transformer action occurs whether the primary field is collapsed or suddenly built up.

The source of energy for this husky stimulation of the primary is stored energy from the capacitor. If you've used a typical electronic photo flash gun, you are familiar with the buildup of power as the capacitor charges until a "ready" light comes on. When you touch the camera-shutter release, a set of contacts operates a trigger circuit and discharges the capacitor almost instantly through a gas-filled tube to create a brilliant flash.

The same thing happens with CD ignition except the capacitor cannot be allowed a prolonged charge time since it must be ready to discharge many times each minute. Fortunately, unlike the batteries in your flash gun, the spinning flywheel magnets provide plenty of the needed energy as they move past an exciter or charging coil. This part of the system is basically a small alternator to provide power to charge the capacitor. A suitable rectifying network changes the alternating current to direct current, and the result is a sort of electrical tank with the capacity to store energy ready to be tapped on each firing requirement.

The timing of this burst of energy into the primary may be done in a number of ways. In some engines an electronic sensing circuit is used to trigger a thyristor much as with the circuits previously discussed. In other engines a small coil—called a *trig-*

Troubleshooting Solid-State Ignition

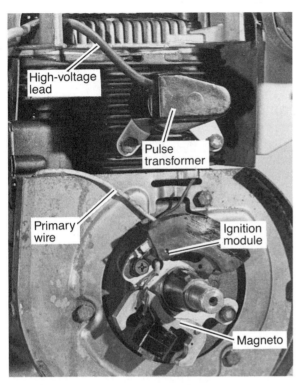

1. Typical solid-state parts include magneto, encapsulated ignition module, pulse transformer, and high-voltage and primary wires.

2. Before assuming solid-state parts are bad, always check for chafing or cuts in the high-voltage lead and primary wires.

3. A close look at this connector clip revealed that it ws bent too close to the bolt head beneath it. Clearances are often tight.

4. Too little clearance was the culprit here, too. These magneto lead wires have begun to abrade on the flywheel hub.

Electronic-ignition kit above consists of a microchip, twist-on connectors, and mounting. The external electronic module mounts as shown below. Place it away from heat.

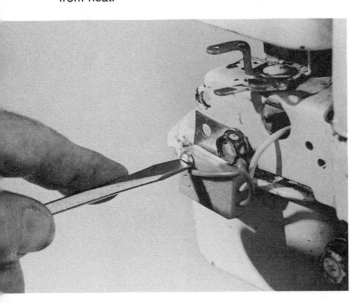

ger coil—is placed near the flywheel, often with a small extension pin to pick up a magnetic pulse from the wheel. As was pointed out, when you have a magnetic field, a coil of wire, and motion, you generate electrical energy. In a CD system, this energy is used to trigger the discharge of the capacitor.

SERVICING SOLID-STATE IGNITION

Solid-state systems—CD or otherwise—are not readily serviced without highly specialized test equipment. In general, the only option is to have a suitably equipped dealer or service representative test the system and replace suspect electronic modules with new parts.

As with breaker-point ignition, a great many electronic-ignition problems are actually simple wire problems. The accompanying photos show the components of a typical solid-state system, as well as basic troubleshooting procedures.

Microchip Replacement Kits

Yours may well be one of the many small engines—including mower engines, chainsaws, hedge trimmers, and the like—on which small microchips the size of postage stamps can be installed to replace the points and condenser. These inexpensive kits are sold at many hardware stores and are available for a number of engine models. The bubble pack they're sold in lists the models they will fit. Each pack also includes explicit instructions and small parts such as screws, wire, and wire connectors needed.

In a typical installation, there is no reason to remove the flywheel or the breaker parts underneath. You simply snip off the primary wire and switch wire where they emerge from the breaker housing, and use twist-on connectors to couple the microchip to the coil. The chip is mounted outside the engine. The entire system is neat, oil-proof and waterproof, and requires no adjustments. Although these electronic systems are relatively new, and are non-factory parts—which I don't usually suggest installing—the microchip units I've tested seem to perform well. Of course, if you already have an electronic system they are not suited for your engine.

11

Troubleshooting and Servicing Electrical Systems

Today, with tractors, recreational vehicles, and even snowblowers being equipped with everything from headlights to electric starters and clutches, even simple and relatively low-power engines are likely to have an alternator. That means servicing and maintaining small gas engines may now entail probing into components that can't really be called engine parts.

ELECTRICAL-SYSTEM COMPONENTS

Among the parts you'll find are batteries and starters, alternators and voltage regulators, interlocks and relays, ignition and safety switches, as well as fuses, diodes, and solenoids. While you can visually detect a leaking oil seal or battered flywheel key, an equivalent inspection of an encapsulated regulator/rectifier, a diode, or even alternator windings requires at least a volt/ohmmeter. Then you must know exactly where to place the probes and what readings you should expect if a part is defective. Moreover, the rapid growth of solid-state technology and the huge variety of systems and components mean specific service information for your particular electrical system is often essential. Random probing will tell you little, and it's possible to damage solid-state parts unknowingly with the wrong techniques.

This chapter describes electrical components, troubleshooting, and servicing techniques common to all small engines to give you a clearer idea of what each part does and how it does it—and help you put the often abstract and involved information found in most manuals to practical and easy use.

Generators

Now obsolete in automobiles, generators have been used widely in small engines in the past and are

sometimes still used, primarily for battery charging. On a small garden tractor, for example, a standard lead/acid battery provides power for starting. To maintain the battery charge, and provide power for lights and perhaps electric implement clutches and lifts, a belt-driven generator mounted on a bracket external to the engine may be used. In many cases the starter switch is set up to feed battery power back to the generator to use it additionally as a starter motor. The belt that drives the generator also serves to crank the engine in this mode.

A volt/ohmmeter is the basic tool for checking electrical systems and associated parts.

This typical motor/generator serves as both a starter and battery charging source. Troubles often involve the mounting and belt drive, particularly when the generator and engine pulleys are out of line.

Such generator units, or *motor/generators,* are relatively heavy and bulky. Instead of compact permanent magnets to provide a magnetic field in which the wire-wound rotor turns, they use wire-wound field coils that serve the same purpose. Remember, we're talking about exactly the same principles that make a magneto work: the movement of either a magnetic field or a wire-wound core, relative to one another. In the generator, a coil of wire wound on an armature moves and the field coils are stationary.

Commutators and brushes. The electrical power developed by the rotating armature must now be transferred to the circuits it feeds. That can be done by sliding contacts, or *brushes,* which ride on a copper ring that rotates with and is an integral part of the armature. The brushes are made of a carbon compound, are self-lubricating, and conduct current well.

If the copper ring were a continuous band, the current would alternate its flow direction as the wound poles on the armature moved past the opposite field coils. That is what happens in an alternating-current generator, but alternating current (AC) will not charge a battery. To change the alternating current to direct current (DC), the copper rings must be interrupted into short segments so that the brushes serve as switches to collect the current from only the set of poles in which the current is flowing the desired direction.

The main points to note are that rectifying the current to direct rather than alternating current is done mechanically by commutator and brushes; the magnetic field is produced by passing current through field coils, not by permanent magnets; and the generator is a separate unit, not part of the engine.

One additional point is important. Since the magnetic field is produced by passing a current through the field coils, it is easy to regulate the strength of this field and therefore the generator output by vary-

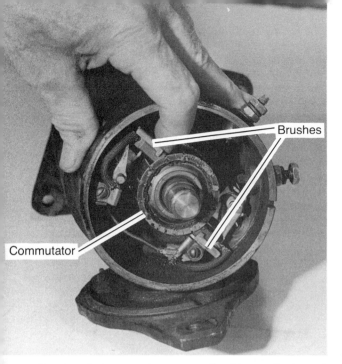

Brushes in a motor/generator should seat firmly and the commutator should be clean and well polished. If not, service is called for.

ing the field current up or down. The relatively small magnetizing current used by the field coils is, in a sense, borrowed from the armature output via the brushes. Most voltage regulators used with generators control the field current to maintain a fairly constant output or diminish it to avoid battery overcharging and burning out lights and other equipment.

Alternators

The alternator, as we know it today, would not have been possible without the invention and development of solid-state devices known as diodes. Diodes take many shapes and forms, sometimes looking like small buttons, though their appearance is unimportant. It is the chemical/electrical makeup of the substance forming the diode that has the remarkable property of passing current one way and blocking it the other way.

Thus, with no mechanical brushes or commutator, an alternating-current output can be changed to direct current for battery charging by passing only one half of the alternating wave form. Visualize the typical flywheel alternator in your engine as having permanent magnets in the flywheel rim, which rotates past fixed wire-wound coils mounted in spoke fashion around the flywheel interior area. The ends of the metal coil armatures are located in close running clearance with the magnets, and as the engine runs, alternating current is generated in those coils. Since the coils are fixed there is no need for a commutator or brushes to transfer current from a moving member to the rest of the system circuitry. Basically, the output of the alternator depends upon the number of coils provided.

Some small alternators in the 3-ampere range are used only for resistance loads, such as lights on small sports cycles and the like, and do not use diodes. The machines have no batteries to charge, and direct current is unneeded. Lights can operate equally well on alternating current.

Alternators used for battery charging, however, must have rectification to provide direct current. The simplest form of rectification, although not the most efficient, is to place a single diode in the alternator output. As the output alternates first one direction and then the other, the diode permits only half of the current to flow. The result is a pulsating direct current suitable for battery charging. By using several diodes set up in what is commonly called a *bridge,* both waves of the alternator's output can be rectified. This is called *full-wave rectification.*

Alternators generate alternating current, which reverses polarity and direction of flow at a frequency reflecting engine speed. A single diode passes current pulses in one direction only—which becomes pulsed DC and will charge a battery. Alternator systems have been the norm in automobiles since the 1960's.

Regulators

Because the alternator uses permanent magnets, and there is no way to vary the strength of their magnetic fields during operation, it follows that alternator output will change with engine speed. It is also true that the charging needs of the battery will vary. A battery that has just been used for starting cranking will be discharged somewhat; until it returns to full charge, the alternator will be called on for a fairly high output. After the battery is recharged the alternator may have a minimal load requirement. By itself, the alternator would have no way of compensating for these changed conditions and would continue to feed power to the battery to the point of boiling and serious damage.

To prevent that, a solid-state device called a voltage regulator senses the load requirements according to the resistance in the battery circuit and automatically establishes a bucking current to reduce output. If your car has an ammeter, you'll see the needle move briefly to the charge side immediately after starting and drift back as the alternator recharges the battery.

Batteries

The battery plays a significant part in the operation of the entire electrical system. If the battery is of the wrong size and internal characteristics, it may be a mismatch for the alternator and regulator. If it has a dead or shorted cell it probably will not charge properly. If the terminals are corroded and not making good contact, charging and voltage regulation will be impaired.

Certain cautions are in order when working with batteries. They apply particularly to modern engines with solid-state ignition and alternator/regulator parts. One major engine builder recommends

This flywheel alternator's multiple coils produce alternating current as the magnets in the flywheel rim revolve around them.

against the use of jumper cables because of the possibility of reversing the polarity through the solid-state parts and causing severe damage. The manufacturer also recommends against the use of electrical welding equipment without first disconnecting the battery. And, as a matter of common sense, it is wise to make certain that the battery, its

By using more than one diode in a rectifier the half-wave rectification can be made into full-wave, which is smoother and suitable for extensive use and battery charging on larger machines.

The solid-state components in this box rectify alternator output and also act as a voltage regulator to prevent battery overcharge.

should provide 40 to 60 starts before recharging is needed. That capacity normally falls off as the battery grows older. Not all NICAD batteries and chargers have the same maintenance instructions and it is important to follow the charging instructions for your specific battery. For example, one engine builder recommends recharging after each engine use, a full recharge requiring about 16 to 18 hours. A common problem occurs when the user plugs in the charger and then forgets to unplug it the next day; prolonged charging is harmful to the battery. The same manufacturer recommends an overnight charge every two months when the equipment is in storage.

Electric Starters

If you're familiar with typical automotive starters, you'll find small engine starters similar in many respects. All are basically DC motors with commutators and brushes, though in some cases they have permanent ceramic magnets instead of wound-wire field coils.

Moreover, some starters are built for a plug-in connection to AC household power and are used on machines without batteries and alternators. That is a common arrangement for snowthrowers, which

cables, and connections and grounds are in good condition before jumping to the conclusion that the engine charging system is in need of invasive troubleshooting. The alternator and charging system are totally independent of the ignition system, even though the parts are located in the same area and one of the coils that appears to be an alternator coil is actually an ignition power-source coil.

NICAD batteries. The batteries previously discussed are standard lead/acid batteries similar to those used in automobiles, though usually smaller because of the lesser demands on them. Another battery widely used on small engines for starting power is the non-acid NICAD type, which provides a limited number of starts and then must be recharged from a 120v outlet using a separate charger.

Engines with NICAD batteries have no alternator-charging capabilities to maintain their charge. That means each successive start diminishes the battery's cranking power. NICAD systems were satisfactory prior to safety rules requiring frequent restarts, after which most mower engines with electric starting began using on-board, lead-acid systems with charging capabilities and much greater repeat-cranking reserve.

Nevertheless, there are still many mowers with NICAD batteries, which under optimal conditions

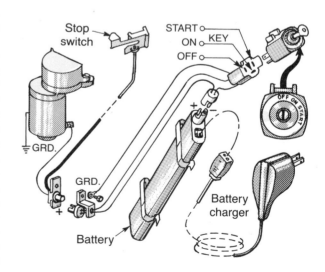

Many walk-behind mowers have used NICAD batteries for starting. Such batteries have a limited number of starts available and must then be recharged.

are often stored for long periods to be called on only now and then for cold starts. With very little running to keep it charged, the battery becomes unreliable and the AC plug-in is more convenient. Note, however, that in most cases the AC power is rectified at the starter and the starter motor is still a DC type, though there are some AC starter motors.

Switches

The ignition and starting switches used on garden tractors and like machines are not always as simple as they appear. And, they can be the source of many mysterious problems that come and go and can sometimes damage the engine's electrical parts.

Remember, the engine manufacturer probably doesn't supply the switch used with your engine. That means the switch may have multiple connections controlling several machine functions, including light and accessory power as well as starting and ignition. Thus, you should never swap an apparently identical switch for the original. Some have as many as six terminals on them, and at least one switch now being used is part of a computer circuit that monitors several engine functions as well as several that are unrelated to the engine.

Remember, too, not to use a switch made for

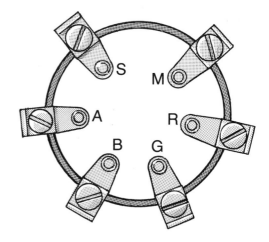

The *S* contact on this multi-contact switch goes to the starter solenoid and should show continuity to the battery when the switch is in "start/crank" position. Magneto switch *M* must show no continuity in *Run* position. *R* = rectifier/regulator, *G* = ground, *B* = battery, and *A* = accessories.

automotive battery-ignition on small engines. On automotive systems, the switch must be *closed* for the engine to operate. On a small engine's magneto system, the switch must be *open*.

Solenoids

If you have ever tried to start your car with a low battery you may have heard a click, clunk, or chatter under the hood even though the starter didn't turn or turned only weakly. The noise was the starter solenoid switch closing.

In some small engine applications, especially on tractors and other large machines, you'll find similar solenoids and often hear similar noises if the battery is low. The solenoid is simply a switch controlled by a small amount of power, which then closes to pass a large amount of power. The reason it's there has to do with the large current draw of the starter motor, which necessitates the use of very heavy wire to conduct that current. Without the starter solenoid, the heavy wire would have to run from the battery, to and through the key switch, and back to the starter, and would also require very heavy electrical contacts in the key switch.

If you disassemble a typical solenoid—sometimes

Typical small-engine starter uses most of the usual parts found on larger automotive starters. The spiral thread meshes the pinion gear with the flywheel.

called a *relay*—you'll find it consists of a wound-wire coil with a hollow core. Inside the core is a loose fitting steel slug, one end with a copper plate or disc riveted to it. A light spring keeps the slug at the back of the coil most of the time, but when you turn the ignition key, power flows through the coil and thus makes it an electromagnet. The slug is then pulled down so the copper plate contacts and bridges two heavy contacts connected to the heavy battery cable and to the starter motor (the solenoid may even be mounted on the starter, as it is in many automobiles). A light current flows through the ignition switch to operate the solenoid. It doesn't take much and that's why even a weak battery will close it and make it clunk. The heavy copper contacts then handle the high amperage draw of the starter.

Fuses

Many small engine charging and electrical systems have one or more small fuses hidden in them. You'll find them located on a small panel under the

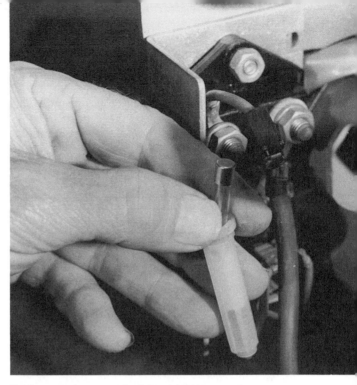

Fuses are where you find them. This one was hiding up next to the starter-switch button.

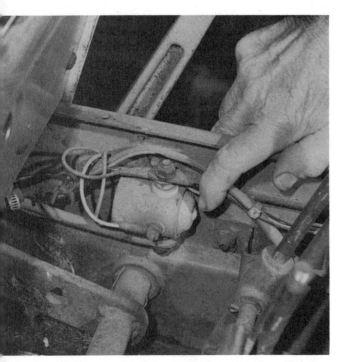

Starter solenoids, like this one buried in a tractor frame, use light current to close a heavy switch and feed power to the starter.

engine blower housing, under a protective cover, or even in a connector in the wiring harness or outlet lead to the battery. Some may not relate to the engine but are there to protect accessories, lights, and the like. Those that do will generally be found close to the engine, since they are provided by the engine builder at the time of manufacture.

Fuses are simple devices to prevent a given circuit from carrying more current than it should. Some of the many reasons why that might happen include a defective device, a shorted or bare wire rubbing on the engine or machine structure, or a malfunction on the input side so that too much power is being delivered. A fuse is usually a thin strip of metal, commonly in a small glass tube with contact caps at each end. When it gets too much current, it melts and breaks the circuit. Sometimes a fuse will fail for no apparent reason, perhaps because of vibration or a hard bump. If a second fuse fails immediately, however, it's a sure sign that some serious troubleshooting is in order before the trouble reveals itself as smoke coming from a place you don't want it to. If you suspect a fuse but it is not readily visible, don't overlook the push-together or twist-together connectors in the engine wiring harness. Fuses are often located in such connectors together with other devices such as diodes.

DIAGNOSTIC TOOLS AND HOW TO USE THEM

As was said earlier, troubleshooting and servicing electrical equipment requires some familiarity with electrical meters—to which the next few paragraphs are devoted. As you put this information to use, you'll find that a great number of electrical difficulties boil down to something as simple and as easily detected as a corroded or loose connection, moisture in a switch, or a non-engine problem such as the safety switch under the seat of your tractor.

Using a Volt/Ohmmeter

A VOM is the best approach to most electrical troubleshooting. This relatively inexpensive device will check for continuity—that what goes in one end of a wire comes out the other—for shorts and grounds, and for the proper opening and closing of a switch. In most cases an ordinary needle-meter

will do fine, although a more elaborate digital read-out is nice if you're really digging into alternator outputs. The power used for simple continuity checks comes from one or two small dry cells in the instrument and is too low to cause damage.

Continuity checks. Let's assume that your electric starter has been erratic and weak. After removing the armature, you place one VOM probe on each commutator bar in turn and hold the other firmly on the armature body. Suddenly the meter needle flicks and you've found continuity—that particular winding in the starter is shorted to the armature body.

Measuring voltage. With the exception of the high-voltage side of the ignition system to the spark plug, the voltages involved in the charging and other circuits are seldom much over battery voltage, or 12 volts. In a few cases you might find as high as 50v in some alternator circuits, but mainly you are dealing with nominal voltages that won't cause a shock or even be felt.

Furthermore, you can switch your voltage/ohm-

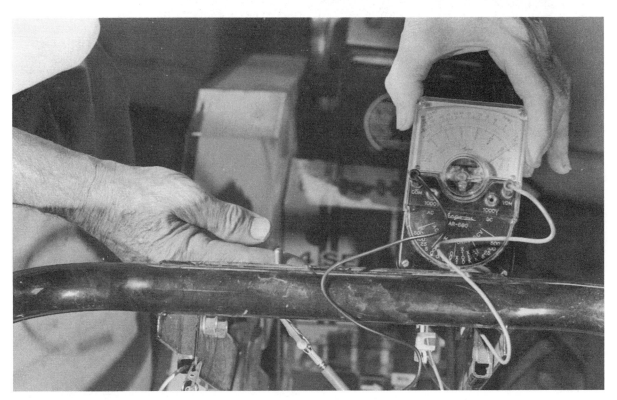

This simple ignition switch on a tiller was improperly connected. A straight VOM continuity check showed the switch was fine but the lead wire was grounded out at the engine.

meter to AC or DC voltage scales as required and measure voltages without any spectacular sparks. That is because most VOMs are rated at an internal resistance of at least 20,000 ohms per volt so very little current goes through the instrument. The high internal resistance also minimizes the effect of the instrument on the circuit being tested.

Voltage is always measured across the two sides of the circuit. If you have a garden tractor, for example, one side of the battery—usually the negative side—will be grounded by a heavy cable or strap connected directly to the tractor chassis or an engine-mount bolt. This means that the physical structure of the tractor and engine is one side of the circuit. The other side is provided by the wiring.

Suppose you put one probe of your voltmeter on the ungrounded or "hot" battery terminal and the

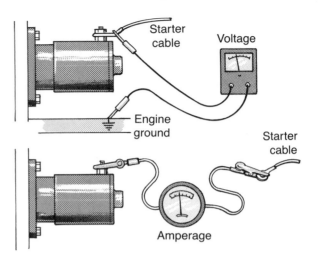

Voltage readings are taken across a circuit, usually between the hot lead to ground. Amperage is checked with the meter in the line.

Here, with the VOM across the battery lead to ground, only about 7 volts showed up when the starter key was turned. The battery terminal posts needed cleaning.

other on a bare spot on the frame or engine. Suppose you read 12v at this point, but when you make the same check with the probes on the starter housing and the input terminal you get only 9v when cranking. This voltage drop tells you that somewhere along the line—maybe at a screw- or plug-in connection or at the solenoid switch—something such as a loose connection, corroded switch contacts, or a poor terminal-post connection at the battery is inserting a resistance in the circuit. By measuring along the circuit at each point of connection you can soon determine where you have 12v and where, on the other side of the connection or switch, you have less. That's your trouble spot.

Measuring ohms. As already mentioned, voltage is electrical pressure and amperage is electrical volume flow. Ohms is the term applied to *resistance* to electrical flow.

Set your VOM on the ohms scale at any value, X1, X100, X1000, or wherever you choose. Touch the two probe ends together and the needle will immediately move to zero because there is no significant resistance between the two bare probe ends touching each other. Now separate the probes and pinch each probe end between the thumb and forefinger of each hand. The needle of this sensitive instrument will move, actually measuring the electrical resistance of your body. If your fingers are dry, the needle may not move very far. Try wetting your fingers to improve the electrical contact; now the needle will jump much farther, perhaps halfway across the scale.

The ohmmeter's sensitivity also makes it very useful for measuring *differences,* even if you do not have specific resistance values. For example, if you suspect that your spark-plug lead is defective, particularly if it is the fragile non-metallic core type, you can measure the resistance of an equal length of new wire compared to the one now on your engine. A high resistance reading or significant difference points to trouble.

Using An Ammeter

Ammeter readings are sometimes given in service manuals, and they generally relate to the battery-charging circuit. Your VOM will probably have a milliammeter scale, but that is not useful in engine power circuits. The reason is that ammeters, which measure current flow, must be installed so that the current in the circuit flows directly through them. Unlike the voltmeter, which measures across the circuit, the ammeter must be able to handle a substantial amount of electrical energy.

The components in the sensitive meter used to measure volts and ohms would have to be much heavier to handle amperes. Alternator output, for example, typically ranges from 3 amperes on small engines to 20 or even 50 amperes on larger engines with heavy electrical requirements. In fact, many ammeters are built with a heavy shunt, or bypass, and sample only a small part of the total energy.

Buy an ammeter if you're interested in thoroughly exploring your small engine's electrical system. Because you'll be measuring both AC out of the alternator and DC out of the rectifier, be sure the ammeter is an AC/DC unit. The fact remains, however, that the volt/ohmmeter—even without an ammeter—is your best electrical troubleshooting tool.

V-belts should have about one-half inch deflection under moderate finger pressure.

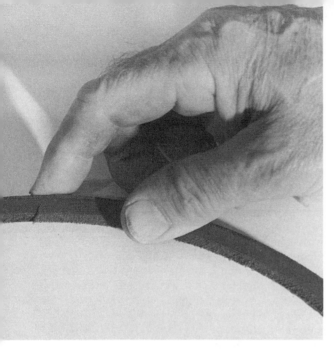

V-belts worn and cracked like the one above must be replaced. The heavy belt load on a motor/generator (below) can also cause the ball bearing at the pulley end of the armature to fail. At least once a season, spin the pulley and feel for roughness. It should spin freely and almost silently.

SERVICING GENERATORS AND MOTOR/GENERATORS

Whether your machine has a generator or, as is often the case, a combined generator/starter motor, you will almost certainly find that it is made by one of the familiar builders in the automotive trade. Should you decide not to service it yourself, you can bring the device to any shop authorized to service its builder's products. Fortunately, however, most troubles with generators and motor/generators tend to be more mechanical than electrical. If you use reasonable care and judgment, you can probably make your own repairs.

Poor starting. The most common problem with motor/generators, for example, is poor starting. The primary reasons are slippage of the belt drive, slower cranking abilities compared to conventional starters, and the somewhat unfavorable pulley ratio between the generator pulley and the engine.

Before deciding that your motor/generator is defective, try pushing your thumb against the outer edge of the drive belt. If you can deflect the belt only about half an inch, the tension is about right. If you can push beyond that, loosen the adjustable bracket that controls belt tension, pry the generator out until the belt is tight, then retighten the bracket. Incidentally, now is the time to replace the belt if it is worn; if your generator has a dual-belt drive, be sure to replace both belts with a matched set.

The belt tension needed to make a motor/generator effective as a starter is also the root of a number of common problems with these units. The high belt load on the drive-end bearing frequently results in bearing failure. The ball bearing is easily replaced, though sometimes the bearing retainer plate inside the generator also loosens and cuts the armature windings.

In addition, motor/generators use brushes that are angled differently than those on conventional generators, apparently because of the unit's dual function. Considered "specials," replacements for those brushes can be hard to find—a concern in heavy garden-tractor service, where such a brush set may last only a few years.

To avoid at least some of those problems it is good practice, perhaps once a season, to relieve the belt tension to inspect and replace the belt if needed. At the same time, spin the generator by hand and listen carefully for any pronounced rattling sounds that

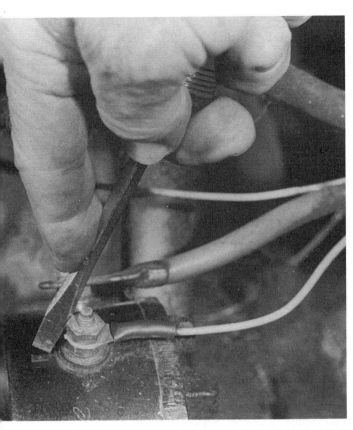

To check the voltage regulator, briefly short the field terminal to the generator housing. If the generator suddenly comes to life, the regulator is defective.

suggest that the bearings may be worn or pitted. Though the bearings are permanently lubricated, the grease tends to dry out and cake, causing lubrication to be lost.

Also check the mounting bracket and adjustment hardware, as well as belt alignment to be certain the belt runs true in the pulley grooves. After retightening the generator, use a VOM to measure starting voltage between the heavy battery-lead connection and ground while cranking. You should be able to read nearly 12 volts if the battery and cable connections are in good condition.

Low charging output. Low charging voltage, usually indicated by a warning light or discharge-side ammeter reading, may be caused by internal trouble in the generator. It may also be caused by a malfunctioning voltage regulator. A quick check can be made by putting the volt/ohmmeter across the

"Bat" lead to ground and momentarily shorting the field terminal of the generator to the housing. If the output on the VOM immediately jumps to 12 to 14v or more, the generator is putting out but the regulator is defective. The machine's ammeter or warning light should also show positive when this is done. The brief shorting to ground temporarily takes the regulator out of operation and the generator gets full field current.

Don't forget to make sure the mounting bolts, spacer washers, and like hardware used to mount and align the generator are snug and tight. Inspect them frequently, since they have a way of loosening. When that happens, the generator bounces and misaligns the belt, and the generator loses charging power.

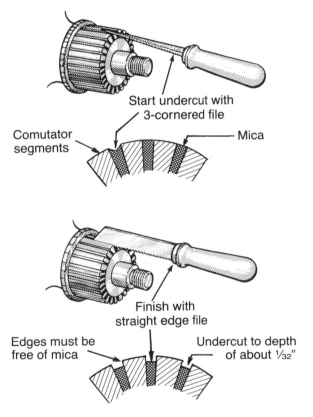

Motor/generator commutator service. If the copper bars are not worn or out of round, try cleaning out the mica as shown and polishing the surface with fine sandpaper.

Unless you have the needed puller to remove the drive pulley and bearing from the generator armature shaft (above), take the unit to a professional. Below: The small screws holding the bearing in place came out in this generator and cut the wires wound on the armature. Repair required soldering in short lengths of wire and coating them with silicone.

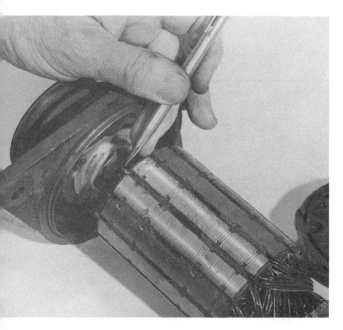

If after the preceding checks the generator still fails to put out full charging power with the engine running at normal usage speed, you can either take it to a repair shop or delve further using the following disassembly and service techniques.

Motor/Generator Disassembly

Although the following procedure addresses motor/generators, the techniques also apply to conventional generators as well as to starter motors. Basic disassembly, bearing and armature checks, and brush service are in most cases very much the same.

Begin by removing the unit from the engine before attempting to open it up for repair. For generators and motor/generators in particular, take special note of any mounting bolts that show signs of battering or worn threads. Look also for any pounding out of the bolt holes in the mounting lugs. If such conditions are severe, it may be well to drill and tap for the next-larger bolt and ream the lugs to fit them.

1. Once the generator is off the engine, clean its outer surface with solvent to prevent getting oil and dirt inside. Also file a small marking notch in the generator housing and at each end plate so you can realign them exactly on reassembly.

2. Loosening and withdrawing the long bolts that pass through the housing lengthwise will usually permit you to tap off the rear cover. *Do this gently and evenly, since the rear bearing will have to come off with the cover.*

3. Listen for a snapping sound at this point as the spring-loaded brushes slip off the end of the commutator. The brushes and their holders will be attached to the cover, which is usually restrained by a wire or wires leading to the field coils. *Do not disturb these.* You may also find a buildup of spacer washers, metal, or fiber on the shaft. Keep these pieces in order and be sure that none are lost.

Checking commutator and brushes. Examine the brushes for length. If they appear short and stubby, they are probably worn and may have been hanging up and making poor contact with the commutator. If they appear of adequate length, check for free movement against the pressure springs by sliding the brushes in the holders. If they are broken or too short you'll have to obtain new ones.

The commutator should appear polished and have a glossy brown color with no signs of arcing or pitting. If it is worn on the brush track a significant amount and if the insulating material between the bars appears to be flush or protruding, the commutator must be turned down and the segments between the bars undercut about 1/32 inch below the surface of the metal. You can do this if you have a metal-cutting lathe or by carefully following the procedure shown; otherwise it is a job for a specialty shop.

Checking generator bearings. Before installing new brushes there are several other checks, and possibly repairs, that should be made. Test the rear for smoothness by turning the inner race with your finger. There should be no trace of roughness, and in most cases the rear bearing will be satisfactory. You can check the front bearing by pushing the end plate out a trifle and spinning the plate on the shaft to listen for sounds of roughness. The front bearing is heavily loaded because of the belt tension and is more likely to be in trouble. Replacing it will, however, require pulling the drive pulley and the bearing from the shaft. The bearing is retained by a thin metal plate, which may damage the armature if the screws loosen. After pressing the new bearing in place on reassembly, thoroughly clean the screws,

holes, and plate and apply a locking agent to the screws.

Checking the armature. You can make a fairly reliable check with a volt/ohmmeter even with the armature still in the housing.

1. Set the VOM on the ohmmeter function and wedge one probe in between the armature metal laminations and the metal pole shoe of a field coil. Be sure you are contacting bare metal on the armature; some have a varnish coating.
2. Carefully run the probe around the commutator bars, contacting each in turn.
3. There must be no flicker of the needle indicating continuity. If there is, the armature is defective and you have a short between a winding and the armature core.

Installing new brushes. Install the new brushes in the holders exactly the way the old ones came out, with particular attention to the routing of the wire leads. The leads are attached with small screws or sometimes with solder. Be sure they're secure and that there is no possibility of the lead wires touching the housing or hanging up the brush in the holder so it can't slide freely.

Installing the rear cover and brushes can be tricky because the brushes are pressed inward by the spring

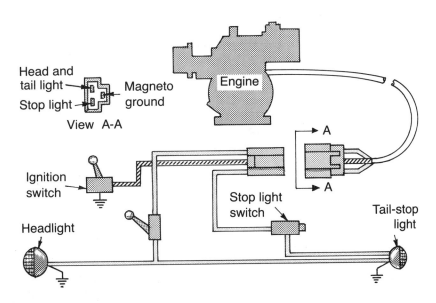

Simplest alternator system has no battery to charge and therefore needs no rectification.

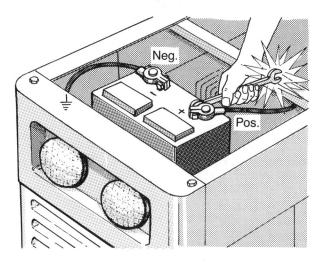

Always disconnect the ground—usually *negative*—side of a battery first. If your wrench happens to touch a metal part, you'll avoid the dangerous arc shown above.

fingers that hold them against the commutator. Thus, they must be held back in the holders and away from the commutator to install the end plate. At the same time, you have to be careful not to let the spacer washers slip off the shaft. Some mechanics make up brush holders, perhaps from a bit of broken starter spring, to hold the brushes until they are slipped partially over the commutator. Another way, for the occasional job, is to withdraw the brushes in the holders so the pressure springs bear against the brushes sideways to hold them in place. Then, with the brushes partially over the commutator, you can work through the gap between the cover and the housing with a small stick or the like to poke the brushes down until the springs snap in place. As a final check, look through the gap with a flashlight to be certain the lead wires are not grounding or in a position to interfere with brush movement.

Polarizing. After reassembling and mounting the generator and being certain that it can be spun freely by hand, it is good practice to *polarize* a generator—or, make sure its internal windings match the rest of the system. This also applies to replacing a regulator. Do this by briefly touching the ends of a jumper wire to the "BAT" terminal of the regulator and the "A," or armature, terminal of the generator. The battery must be connected. Only the shortest flash is needed; do it quickly to avoid damage.

If after the preceding checks and service procedures the generator still fails to develop output—and the regulator is not defective—the problem may be burned armature windings, shorts or discontinuities in the field coils, or internal shorts in the armature. Such problems require professional service and equipment to locate and correct.

SERVICING ALTERNATORS

The generators and motor/generators just discussed presented a number of mechanical problems with mounts, bearings, brushes, and commutators, which were relatively easy to detect visually and repair. Alternators built into small engines are generally free of such concerns, since the only bearing involved is the crankshaft bearing and there are no brushes or commutators. Thus there is little to go physically wrong with an alternator with the possible exception of a misrouted wire rubbing on the flywheel. That means for most troubleshooting you'll be using a VOM.

The usual indications of trouble in the alternator system are a dead battery or lights that are dim or not receiving power at all. The obvious place to start is at the battery to determine its condition and the quality of the cable connections to the terminals. If the system is for lighting only, start with the lights

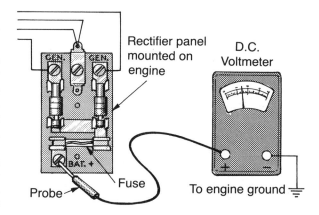

Checking an engine-mounted rectifier panel. If there is no output with the engine running and the battery disconnected, check for input to the fuse. Then check the output of each diode. If diodes pass current both ways or neither way, replace them.

and their receptacles. A continuity check with a VOM across the bulb terminals will tell you if the bulb has a broken filament. A voltage check at the receptacle terminals will tell you if power is reaching the lights.

Above all, in making these electrical checks *be careful!* Many of them require that the engine be running at close to normal usage speeds. All of the previous precautions about loose clothing, hot parts, and placement of your hands and feet apply here. In addition, before starting, remove all rings, wrist watches, and bracelets. Although your small engine's electrical system does not produce lethal voltage, a short across a 12-volt battery through a metal ring can melt the ring instantly with serious consequences for your finger. And never use metal-handled screwdrivers, pliers, or other tools without handle insulation when servicing live circuits. They're just too easy to short to ground.

For the most part, you'll use the AC setting on your VOM. Remember, however, that your small engine alternator output is AC until it passes through a diode rectifier. If your alternator is supplying power for lighting only, the AC scale should be used. Some alternators have split outputs with AC for lighting and DC for battery charging. The rule is to use the AC setting for all readings on the engine side of the rectifier; also use the AC setting if no rectifier is used. Use the DC setting on all circuits from the rectifier to the battery, and from the battery to accessory circuits.

Checking battery connections. In addition to the cautions for electrical work previously described, connecting and disconnecting the battery warrants some extra precautions. Always begin by disconnecting the ground side—the side that's grounded to the engine or machine frame. If you do not do this, sooner or later you'll bridge between the hot, usually positive, terminal and some part of the engine or chassis with the wrench. The result will be an intense arc that will damage the wrench, probably burn you, and possibly cause a battery explosion of sulphuric acid. Remember: *Disconnect the ground side first, reconnect it last.*

The nature of solid-state parts makes them vulnerable to reversed polarity. If your battery connections are not clearly distinguishable, use red tape, nail polish, or the like to plainly mark the positive cable terminal and the battery terminal to avoid reversing the connections.

Also, since the alternator circuits and the ignition

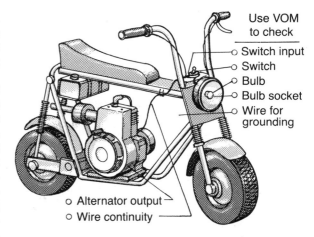

A lighting failure on a minibike suggests an upstream check with a VOM. First check the bulbs; unregulated power can easily burn them out.

circuits are independent, the engine will run even though you disconnect the battery or the battery-charging circuit. When making tests, make them quickly, then stop the engine and reconnect the battery and charging circuit. Do not run more than a minute or so with the battery disconnected. Even if the battery is dead or shorted, whatever electrolyte remains will prevent the rectifier and regulator solid-state components from overheating. Finally, if you plan to do any electric welding on your machine, disconnect the battery and rectifier regulator first.

Checking other connections. In many cases you'll have to use test probes on the alternator output connections or other wires in or near the engine flywheel. Sometimes you'll have to remove the blower housing, route the wires outside for access, and reinstall the housing to run the engine. Never run the engine without the blower housing, since this will inevitably cause serious overheating and damage. Take particular care to observe exactly how the wires were originally routed. Then, carefully route them outside so they won't contact the spinning flywheel or ring gear. Also make sure that movement or tugging won't drag them into contact and that they aren't being cut or crushed when you replace the blower housing.

If the wires are extremely short, it's safer to rig some form of extension leads for your test probe contacts rather than disturb the original routing of the wires. It is also possible to seriously damage the alternator if a loose, bare wire end is allowed to contact the engine structure to ground, or if two wires happen to short together. *Use electrician's tape to temporarily cover all bare contacts.*

Troubleshooting Alternator Circuits

There are two possible orderly approaches to establishing whether the circuits fed by the alternator are sound. One might be thought of as working upstream to the alternator—the source of power—and the other as working downstream.

An upstream check toward the alternator not only weeds out easy problems, it establishes any cause-and-effect relationships—such as defective wiring—that may be responsible for alternator trouble. An example of an upstream check might involve a minibike with lighting circuits only and no battery charging. Since there is no battery, begin by making sure the machine is securely parked, preferably with the wheel-drive removed to prevent accident since the engine must be running fast enough to duplicate normal operation.

1. After first checking the lights for burnout, check the light receptacles. Because lighting circuits are usually grounded on one side to the machine frame and use a single wire running to each light, your readings will be between the accessible terminals of that wire and the machine frame.
2. Next check where the wire enters and exits the light switch. If the switch is defective, you may have power in but none out.
3. If there is no power to the switch, shut off the engine and disconnect the feed wire to the switch at the connector plug.
4. Test the feed wire for continuity to be sure it isn't grounded or broken. Do this using the ohms setting between the frame and the switch input terminal. A reading here shows the wire is bare somewhere along the line and is shorting to the frame.
5. If the wire is not shorted, test it for continuity by touching the VOM probes to the terminals at each end. The meter should show zero resistance. If it doesn't, the wire is broken.

If the circuits are defective there may be nothing wrong with the alternator, but a fuse in the output may have burned out because of the short to ground. With the engine still running at medium speed, take

Connecting VOM probes to the two pins in the harness connection shows that the windings in the alternator coils have continuity. Not all alternators are wound this way.

INTERNALLY GROUNDED ALTERNATOR

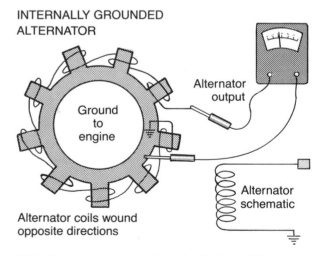

This alternator has a single output wire and the opposite end of the wire is grounded to the alternator stator. A VOM check between the output wire and ground should show continuity.

an AC voltage reading by probing between one of the alternator output-plug connections and ground. You should find an output of 8 to 10 volts AC. If not, the alternator, a fuse, or the alternator output wire is defective.

If in checking at the alternator output plug you find three connections or prong sockets, one is the ignition switch wire and the other two are lighting circuits. Be certain you're not trying to get an alternator reading from the magneto. Remember, too, that because your small engine has magneto ignition it will start and run just fine with the ignition-switch connection open. To stop the engine, however, you'll have to short out the spark plug or close the choke.

Internal Alternator Checks

To locate possible defects in the alternator you will probably have to remove the blower housing. Carefully inspect the wire leads from the alternator for signs of rubbing, abrasion, cutting, or other conditions that would ground out alternator output. You can then use your VOM to make two simple checks of the alternator windings: for continuity and for internal grounding.

Your alternator windings may be either of two different configurations. In one configuration the wire to which you have access at one end is wound around a series of several alternator coils and grounded to the frame of the stator—and thus the engine—at the other end. If you put your VOM on the ohms setting, putting one probe in the alternator output lead and touching the other probe to ground on the engine, you should get a solid indication of continuity. If you don't, the wire is broken somewhere. If you remove the flywheel, following the procedure on page 112, you may be able to locate the break.

What the preceding test doesn't tell you, even if you do get a continuity reading, is whether the wire is simply normally grounded at the opposite end or is shorted to ground somewhere in the coil windings. Again, if you remove the flywheel and disconnect the ground end of the wire from the stator, you should now get no continuity to ground but should get perfect continuity through the wire end to end. Some ground-end connections are relatively easy to open, while others are very difficult.

The second configuration of alternator winding is sometimes termed a "floating" circuit in that neither end is grounded. You can guess at this if examina-

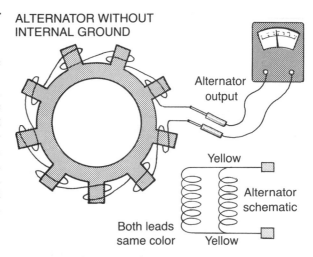

ALTERNATOR WITHOUT INTERNAL GROUND

Alternator output

Yellow

Alternator schematic

Both leads same color

Yellow

On alternators not grounded internally, both ends of the wire are brought out. There should be continuity between the two ends but no continuity between either end and ground. If there is, the coils or lead wires are shorted out.

tion of the connecter plug shows two yellow wires, for example, each terminating in a prong connection with the ignition switch connection in the center. Here, you are almost certainly looking at both ends of the same wire. Touching the VOM probes to each of these terminals should show continuity and confirm that this is a single wire going in, ~~~~~ the coils, and out. Knowing that, you clear~~~~ould not be able to put a probe in either terminal and show continuity to the engine ground structure. If you find continuity, the coils are grounded internally somewhere.

Although there are certainly more complex procedures, using special-service test equipment, the preceding steps should yield worthwhile results and be well within any home mechanic's capabilities.

Solid-State Troubleshooting

The general test sequence suggested for a simple minibike alternator system applies equally to more complex systems using solid-state parts for a DC battery charging system. *Remember to set your voltmeter to DC mode.* It is likely that you'll now have a multiple-connection starting switch in the system, and it bears careful checking before assuming your

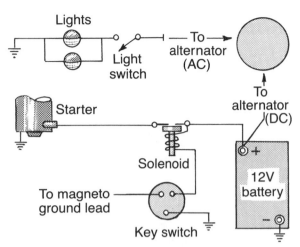

INSULATED SOLENOID WITH KEY SWITCH

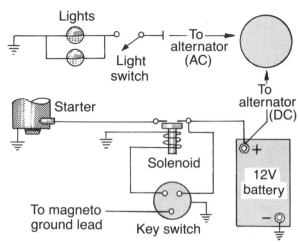

GROUND SOLENOID WITH KEY SWITCH

Some alternators have two or even three circuits to handle various accessory loads. Basically, a DC bat-tery-charging circuit is provided and another AC light-ing circuit is used for lights only.

alternator or rectifier solid-state parts are defective. A wiring diagram is useful at this point, although a little hard thinking and some continuity checks of the switch in various positions will usually solve the puzzle.

Checking Unregulated Systems

The simplest solid-state system is unregulated and consists of one or two rectifying diodes and usually a fuse. These parts may be located on a small panel on the engine, perhaps under the blower housing, or simply in a connector plug between the engine and the rest of the circuits.

1. If the battery is not charging, and you can find nothing wrong with the switch or the wiring to the battery—with the engine running at medium speed, take a DC voltage reading to ground at the point where the charging wire leaves the rectifier.
2. If the wire doesn't terminate at a panel where it is easy to probe, a pin through the wire insulation will serve as a contact.
3. If no voltage is found there, stop the engine and check the fuse in the panel or connector for continuity. Sometimes a fuse looks good but is defective.

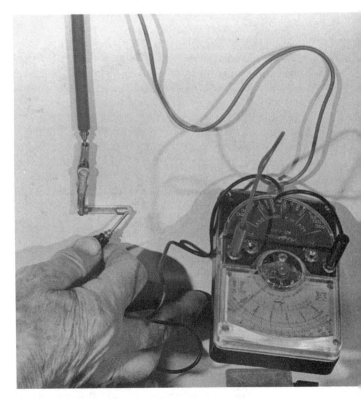

A defective fuse may sometimes appear good. A continuity check with your VOM will tell you for sure.

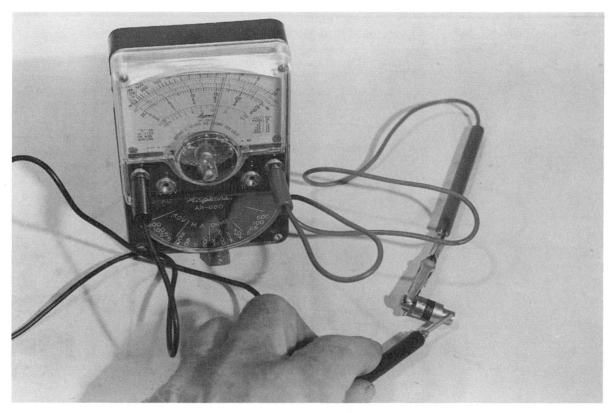

Most diodes have an arrow showing which direction current should pass. If current passes in both directions or not at all, you can be sure that the diode is defective and must be replaced for the system to work.

Replace the fuse with one of the same rating, if necessary, but don't overlook the principle that when a fuse blows it usually means an overload. You may well have a short in the machine.

Checking diodes The diodes are easily checked with the ohmmeter by touching the probes to each end and then reversing the contacts. A healthy diode should pass current one way but not the other. If it doesn't pass current at all, or passes it both directions, it must be replaced.

In many cases a defective diode must be snipped out of the wire in which it is installed and a new one soldered in place. To do this:

1. Strip back enough insulation on each side of the gap left when you cut out the diode to allow bending a small hook. Form like hooks on the diode wires.

2. Make a final check to be certain current flow is in the right direction, in this case away from the engine. An arrow or mark

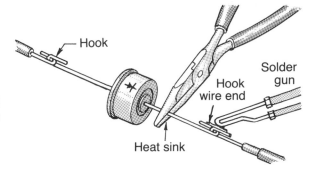

Diodes are commonly replaced by cutting the defective one out of the line and soldering in a new one. Be sure to prevent soldering heat from reaching the diode.

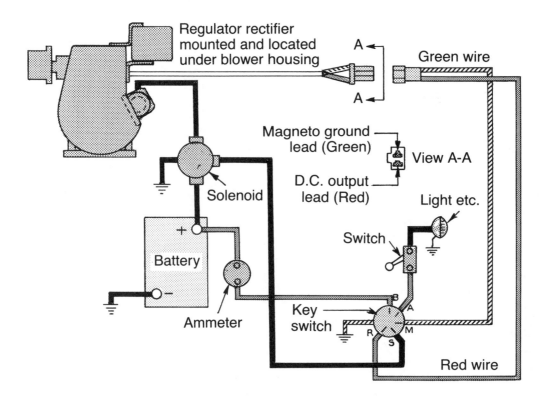

Output of the 7-ampere alternator on this engine is fully regulated and all-DC.

on the diode usually shows current-flow direction, though it doesn't hurt to check with a VOM. If you have some heat-shrink tubing, slip a section over each wire before making the hookup.

3. Next, hook the wires together at each end of the diode and squeeze down on the joints with pliers.

4. Remember, all solid-state devices are vulnerable to heat. To protect the diode when soldering, grasp the wire between each solder joint and the diode with the pliers to block the flow of heat to the diode. *Use resin-core solder—never acid-core or flux.* Be sure to allow each joint to cool before removing the pliers or moving the wire.

If you used the heat-shrink tubing mentioned in Step 2, you can now slide it over the joints and heat it for a tight seal. Or, protect the wires and diode with electrical tape.

Checking regulator-rectifiers. When the rectifier diodes and a solid-state voltage regulator are combined in a single sealed unit, the same possibilities for trouble are present as in the systems previously described. Now, however, there should be differences between the voltages read on the AC input side and the DC output side. That is the action of the regulator—to reduce the alternator output and stabilize it for battery charging. Though you'll need to know the specific voltages for your particular engine, the basic check here is a downstream check starting at the alternator to determine if it is delivering the specified voltages.

If the procedure on page 162 yields a satisfactory alternator output, you can in many cases check the regulator output to the charging circuit. In some engines the system must be connected to a battery with a charge of at least 5v for the regulator to operate properly. A satisfactory reading out of the regulator points to problems in the wiring system or switches, which must then be checked as previously described. Note that the regulator must be grounded physically to the engine or chassis. That means tests made with the regulator off the engine require a jumper wire between the regulator housing and the engine.

SERVICING MECHANICAL VOLTAGE REGULATORS

If you happen to have a small engine with a belt-driven generator or motor-generator, you'll find a small black box either mounted directly on the generator or on the machine frame. In the days when this type of unit was used on every automobile, if the ammeter needle on the dash dropped over to a disheartening discharge it was common practice for one person to watch the ammeter while another lifted the hood and whacked the box sharply with a screwdriver handle. If the needle jumped back to charge, even momentarily, it usually meant that the generator was putting out but the voltage regulator was faulty.

The reason that procedure worked is that instead of mysterious solid-state devices regulating voltage, equally mysterious little coils, springs, and contact points did the job. The screwdriver jolt momentarily freed up a sticking contact point or otherwise disturbed the vibrating contacts so something happened that showed up on the ammeter.

Regulator components. Remove the cover from a mechanical regulator and you will see two coils. One is a cut-out coil. Its purpose is to open the circuit between the battery and the generator when the engine is idling slowly or not running, thus preventing a backflow of current that would discharge the battery. With the engine above idle speed, power from the generator energizes an electromagnetic coil which pulls down a moveable arm, or armature, to close a set of contacts so power can flow to the

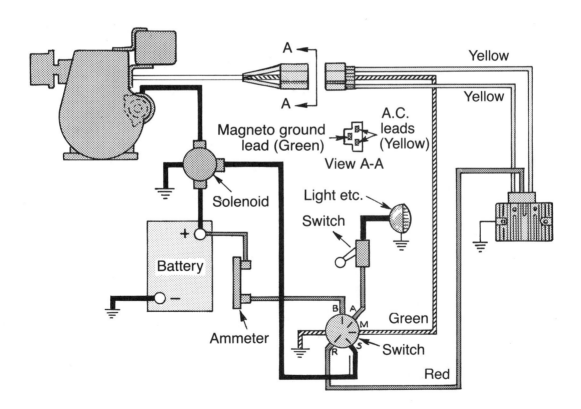

This externally mounted regulator/rectifier must be grounded to the engine to function and for testing. All the power for such a system is supplied from the battery, as can be seen here.

Testing External-Mount
Regulator/Rectifiers

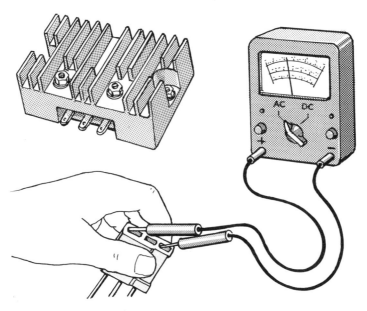

1. Check the input from the alternator with the VOM switch in the AC position and both probes connected across the yellow output leads. Failure to produce the specified voltage suggests that your trouble is either a defective alternator or damaged output leads.

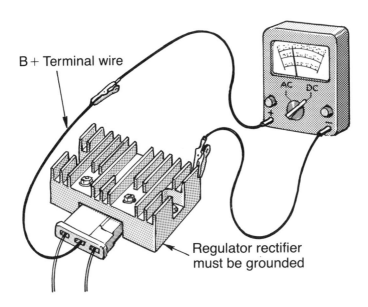

B + Terminal wire

Regulator rectifier
must be grounded

2. With the regulator grounded and the engine running between 2,500 and 3,600 rpm, regulated DC output from the center wire at the regulator should be 13 to 20 volts for this particular unit. *Check a manual for the specs for your particular engine or machine.* Specifications often vary by make or even model.

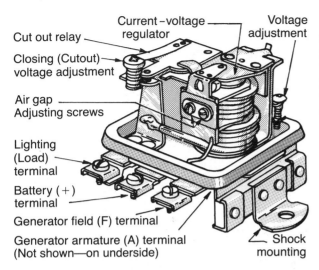

Cut out relay

Current-voltage regulator

Voltage adjustment

Closing (Cutout) voltage adjustment

Air gap Adjusting screws

Lighting (Load) terminal

Battery (+) terminal

Generator field (F) terminal

Generator armature (A) terminal (Not shown—on underside)

Shock mounting

A two-unit voltage regulator commonly used on automotive-type generators and motor/generators.

battery. Once closed, it should stay closed until the engine stops or slows down so the generator output voltage is less than battery voltage.

The other coil regulates voltage and current. If you watched the voltage regulator at work you'd see a small, twinkling arc between the contact points as the armature moved rapidly up and down to open and close the contact points. The rapid vibration results from the opposing forces of a small spring on the back of the armature and the magnetic action of the coil. If you remove the regulator from its mount and look at the bottom you'll find a length of insulating material wound with resistance wire. The contacts are actually connecting and disconnecting this resistance in the generator field-coil circuit. If the field coil current was to run unchecked, the excess magnetism would cause the generator to put out excess current and too high a voltage, thus damaging the battery and generator. By putting a resistance of just the right amount in the field circuit, the output is kept in the proper range.

If too much resistance is in the field circuit, the magnetism will be too low and the output will not be enough to charge a battery. In short, the entire rather delicate mechanism is carefully balanced so the gap between the contact points and between the armature and magnet coil, as well as the spring tension, all balance out to hold the voltage at the desired level.

Cleaning the points. Remember that mechanical regulators were originally built for extended automotive service and are not likely to be worn out in the relatively short hours most small engines run. A problem is more likely to reflect the fact that your tractor or other machine may see only intermittent service and be subject to corrosion and oxidation.

1. With the battery disconnected, begin by looking for any signs of corrosion or oxidation on the contact points. Do this very gently to avoid bending the delicate parts or stretching the springs.
2. If necessary, clean the points using a thin, fine point file made especially for that purpose.
3. Gently press the contacts together, making sure the armatures are working freely and no dust or debris is present. There's an excellent chance that when you start your engine, your regulator will again be working.

Further adjustment requires specific information on the contact gaps and clearances, as well as some

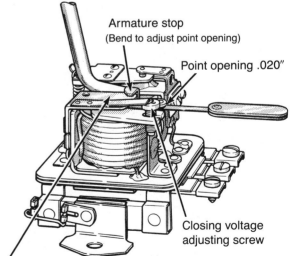

Armature stop (Bend to adjust point opening)

Point opening .020″

Closing voltage adjusting screw

Armature (Raise or lower to adjust air gap) Air gap .020″ (Armature to core)

Clean cutout-relay contact points with a fine file and make adjustments only if you have the proper specs for your regulator.

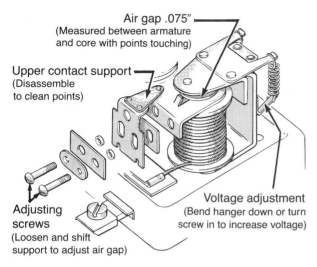

Air gap .075″
(Measured between armature
and core with points touching)

Upper contact support
(Disassemble
to clean points)

Adjusting
screws
(Loosen and shift
support to adjust air gap)

Voltage adjustment
(Bend hanger down or turn
screw in to increase voltage)

Contact points on the current/voltage regulator part of the unit must also be clean. Use a fine file and adjust gap as required, then run the engine and use a VOM to adjust output.

delicate tweaks on several adjustment mechanisms you'll find inside. Unless you have experience at this and have equipment to read voltages accurately, you're better off taking the unit to an automotive electrical-repair shop. There are literally thousands of variations on such units and you'll need the correct specification for any given series.

TROUBLESHOOTING STARTERS

When your electric starter shows no sign of life, don't assume that it's the starter's fault. Before the starter can work, it must receive power from the battery. Review the section on troubleshooting the ignition system, Chapter Five, particularly with reference to safety and starting switches. On mowers with safety brakes, be certain that the starter interlock switch is closing and the stop switch is opening. If a VOM reading at the starter input terminal doesn't show voltage with the switch or key in the

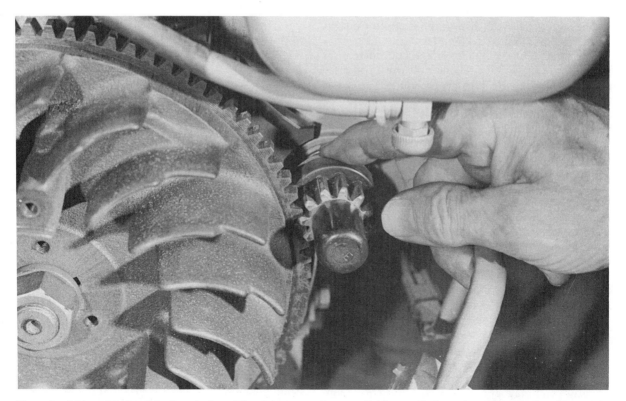

You should be able to slide the starter pinion into engagement with light finger pressure, and there should

be no stickiness. If there is, clean starter-drive parts with mild solvent *but do not oil*.

Checking Electric Starters

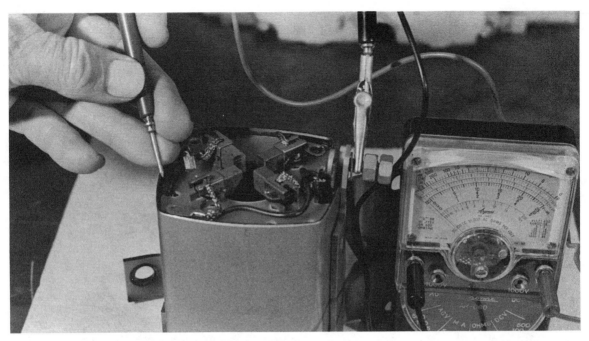

1. To be sure the field coils aren't grounded, clip one VOM probe to the power-input terminal and touch the other probe to the starter body. There should be no continuity.

2. Check armature windings by clipping one probe onto the armature laminations and touching the other one to each commutator bar in turn. Again, there should be no continuity.

start position, something is wrong downstream and a VOM check such as the one described for the alternator is needed.

If adequate voltage is reaching the starter, expect to hear at least an indication that it is trying to start the engine. Disconnect the spark-plug lead and try to crank the engine by hand. If it feels locked up or hard to turn, check for a machine clutch that is engaged or for something fouling a mower blade.

If voltage is reaching the starter and the engine is free to turn, but you hear nothing, the trouble is probably inside the starter. It may be a brush not making contact or you may have an oil-fouled commutator. A burned-out or shorted armature or field coil is also a possibility. The only way to determine the cause is to disassemble the starter and use your VOM, plus visual inspection in the same manner as described on page 158 for the motor-generator.

When a starter cranks slowly, the battery may be at fault, but you may also have brush trouble such as wear, weak springs, or a dirty commutator. It is also possible that a starter bearing is worn and the armature is dragging on the field poles.

Finally, you may have a starter that spins vigor- ously but does not crank the engine. That calls for removal and inspection of the pinion and engaging device for sticking, dirt fouling, or breakage. To check the starter drive:

1. Remove the blower housing for good visibility, if necessary, and disconnect the spark-plug wire for safety.
2. Energize the starter and note whether the pinion moves into engagement with the ring gear.
3. Some starters have a fiber-washer clutch under the pinion designed to slip under heavy loads such as backfire. The clutch should *not* slip in normal cranking. If it does, it must be replaced.

In most cases, however, the spiral-driven pinion will be sticking and not moving into mesh with the ring gear. Although there are many variations on this mechanism, most are relatively easy to disassemble by inspection. Use a mild solvent to remove dirt and gum from these parts and dry them thoroughly before reassembly. Do not use oil on the starter drive parts.

OVERHAUL OR REPLACEMENT

12

Should You Rebuild Your Engine?

The previous chapters of this book were intended to give you some general insight into what makes an engine run, and what it takes to perform minor repairs and keep a relatively healthy engine in good condition. Most of the procedures described—with the exception of head-gasket replacement, which requires a torque wrench—can be completed with ordinary tools.

Repairs that require removing the valves, piston and connecting rod, or the crankshaft require more tools. They also necessitate some value judgments as to whether such work is worthwhile in terms of time, money, and the useful life of the equipment the engine powers. In many cases you will be able to reach a certain level of repair and then have to call on a professional shop for work you can't do at home. Examples might be oversizing a cylinder bore or installing new sleeve bearings and reaming them to size.

A NEW SHORT BLOCK—AN ALTERNATIVE TO REBUILDING

If rebuilding your engine will entail a lot of costly machine-shop work, it almost always pays to investigate buying a factory-new short block. A new short block ordinarily includes the cylinder and crankcase, complete with piston, valves, seals, bearings, etc., exactly the way your original engine was built. You then remove the cylinder head, blower housing, flywheel, carburetor, and miscellaneous parts from the old engine and install them on the new block. In my own experience managing field service for an engine builder, it was usually much less costly to start with a short block because the labor time of new parts going together on an assembly line was vastly less than the man-hours a mechanic would expend in disassembling, cleaning, inspecting, repairing, and reassembling replacement parts.

That logic doesn't necessarily apply if the engine involved is from your lawn mower and you have plenty of long winter evenings to putter along rebuilding it. Although rebuilding a simple one-cylinder engine may require only two or three hours according to the dealer's flat-rate book—and some estimates are as low as 45 minutes—be assured that you'll need much more time, especially if you've never done the job before. Besides, trying to hurry takes all of the fun out of the job. If you enjoy working with mechanical things, engine rebuilding is fun and a craftsmanlike job is something to be proud of.

Thus, if getting your machine back into action is urgent and you can't take the time or don't have the confidence or the tools for rebuilding, the short-block route is the way to go. Or, if you've made at least a partial disassembly and determined which parts will need replacement, you may find that you can get a factory-assembled short block for only a small amount more. That should put you back and running with about one hour of work.

DOING IT YOURSELF

If some of the preceding sounds a bit negative towards overhauling your own engine, I don't intend it to be. In most cases, unless the engine and driven equipment are both basket cases beyond repairing so they are safe to operate, I'd say go for it. Here are my reasons:

- You don't have to pay mechanic's wages or the owner's overhead and profit. Those are the most costly parts of mechanical work.
- Small-engine parts are not as costly as auto-engine parts.
- Even if you plan to trade in the machine, you'll do much better with a good-looking, smooth-

Shown above are the parts you remove from your old engine and replace on the short block (below). A new short block includes all the vital wearing parts, factory-fresh and ready to go.

running engine than you will if you present it and say it doesn't run and you don't know why.

- The tools you'll use will last a lifetime and you'll probably use them many times.
- The experience and satisfaction of turning a clunker into a smooth-running engine is one of life's more gratifying experiences.

Before you even start to disassemble an engine for repair, however, be sure to think through why it's necessary. Some reasons include:

- Extremely hard starting, either because of ignition-component failure or lack of compression due to stuck or leaking valves.
- Serious power loss, perhaps coupled with very high oil consumption, both of which suggest stuck or worn rings or a worn cylinder.
- Sharp knocking or rapping noises, which may be bearings about to fail completely, a loose or damaged piston, worn valve gear or camshaft drive, or a loose flywheel.
- Overheating, even seizing up or becoming hard to turn. May be caused by lubrication failure, a lost or damaged oil pump or connecting-rod dipper, or a plugged oil passage to a bearing.
- Constant oil loss around the crankshaft. May be a combination of worn crankshaft bearings and worn oil seals. On a two-stroke the engine may not run or be very hard to start.
- Rough running and shaking that cannot be traced to off-balance driven parts such as a mower blade, or to loose engine mounting. Could be a loose or damaged flywheel.
- Grinding, rough, or squeaky noises when you turn the engine by hand. Often caused by rough or damaged ball or roller bearings.
- Obvious cracks in the flywheel, cylinder head, or crankcase.
- Flutter or wobble in a crankshaft driven part such as a mower blade or drive pulley, which tells you that the crankshaft is bent.
- Damaged keyway grooves in the crankshaft drive or magneto drive end.

You may recall the histories of some of those conditions. For instance, you may vaguely remember hitting a stump with your rotary-mower blade, or perhaps allowing the engine to run low on oil. Perhaps you noticed that the last few times you ran the engine there was a hint of popping and crackling in the exhaust. Maybe you've noticed that every

Oily deposits blowing out the exhaust of this tiller engine leave little question that the piston, rings, and cylinder are badly worn.

Leaking lower oil seal on this two-stroke will cause poor crankcase compression and hard starting, as well as weak running.

Corrosion and probably low oil caused this crankshaft roller bearing failure.

There is no real cure for a battered crankshaft keyway such as this. The shaft needs replacement.

time you added oil the engine took a little more, or that you're getting more blue smoke in the exhaust and a dirty, oily deposit is building up around the exhaust area.

Diagnosing The Problem

Your first step before getting into an overhaul is to disconnect the spark-plug lead and gently rotate the crankshaft. Rock the shaft back and forth in several positions. Any rattle or feel of impediment is a sure sign of internal failure. If the shaft can be rotated freely with no feel of compression, it's a fair bet that the connecting rod, crankshaft, or piston is broken. A broken connecting rod may or may not destroy the engine. If it fractures the crankcase or comes through the side, however, you may as well forget rebuilding at home. While a skilled welder might weld or braze up the crankcase and somehow keep the bearings in line, the cost is likely to be outlandish and seldom worthwhile, unless the engine is some sort of collector's item.

Strain the oil. If the engine can be rotated and feels fairly normal, try straining the oil through something such as a scrap of Nylon stocking. Bright bits of metal filtered out of the oil are sure signs of serious trouble. Those may be bits of piston, bearings, cylinder walls, or crankshaft. A magnet will

show if they're cast iron or steel. Aluminum and non-ferrous metal will not pick up. Of course, if your engine has an aluminum block with no liner, the aluminum particles may be from the block itself.

On the other hand, if a valve head has snapped off and battered its way through a piston, it's entirely possible that the damage is confined to the piston and cylinder head, and the repair is fairly simple and inexpensive. You may be able to spot this by lack of compression.

Remove the spark plug. Almost certainly, any loose part in the combustion chamber will batter the plug tip badly. Looking through the plug hole with a small pen light may give you further evidence, and pressing your thumb over the hole while turning the crankshaft will reveal if there's any sign of compression. If you've already found metal particles in the oil and there's no compression, expect a broken valve or a hole in the piston. The latter might be detonation damage. If there is no compression but you did *not* find metal particles, you may have lucked out with nothing worse than a stuck valve.

In all of the preceding, what you're really trying to determine is the seriousness of the condition. Ordinary wear, bad seals or gaskets, valves that need replacement or grinding, worn piston rings, or faulty ignition parts all come under routine service and are well within the scope of most home repair efforts. Only inspection will tell.

Disassembly of this engine shows a broken connecting rod, camshaft, and lifters; a scored piston and crank-shaft; and a hole in the crankcase. The cause was running without oil.

A crankshaft bent from hitting a stump is a lesson in itself. So is scoring and seizing from running without oil. And a burned exhaust valve or heat-warped head should enlighten you about keeping the debris out of the cooling fins. Some other failures are also easily traced. An oil dipper that broke off the end of a connecting rod is a sure cause of lubrication failure, while a broken-off valve points to excessive tappet clearance. But other than such fairly clear cause-and-effect situations, the answer to what went wrong may be a matter of guesswork.

Service Bulletins And Factory Specs

Issued by engine manufacturers from time to time, service bulletins are sometimes the result of reports from the field regarding, for example, an adjustment procedure on a carburetor or governor. Other times a given part or assembly develops a record of rapid wear or poor performance, and a new part or modification is recommended. Typically such bulletins describe the trouble and advise the dealer of the new part number and how to identify the old and new parts.

If you can find a dealer who will let you scan through the bulletins applicable to your engine, by all means do so before attempting a repair job. If your engine broke a connecting rod and you discover a bulletin advising dealers to replace that part with an updated rod, you can be fairly sure the engine maker knows the trouble. Whether you can press a claim on it is another matter. The same applies to other parts that have shown up as vulnerable often enough to justify a service bulletin. If nothing else you may be able to sooth your feelings by talking the dealer or field representative into a rock-bottom price on new parts and gaskets.

In such cases, always be sure to get the new part number. Updated valves, lifters, or camshafts that look exactly like the old ones may have been subjected to different hardening or heat treating, or may

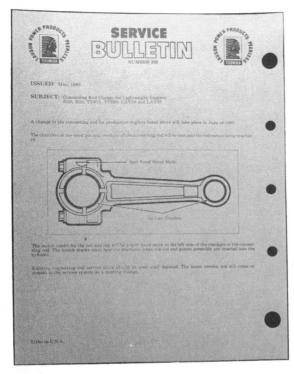

Your dealer's service bulletins show ongoing changes in parts, techniques, and adjustments, in addition to other useful information.

be of a slightly different alloy. Simply eyeballing the parts will seldom tell you anything.

Also—because of the sheer number of small gas engines out there—serious engine work is another area where you'll need a shop manual that provides the vital fits, clearances, and torque values specific to your particular engine make and model. Check your local library or try contacting the engine manufacturer directly.

How To Judge Wear

To gain an image of what wear really is, assume that you have an automobile of the junker variety. The engine uses oil, the body and suspension rattle, and the drive line sounds terrible. Actually, all that separates the car from "like new" is about 3 or 4 ounces of metal that has been worn away over the years. Perhaps an ounce, probably less, has actually been worn away in the engine. In a small gas engine,

the worn metal might not fill half a teaspoon but the engine is essentially worn out.

Wear occurs at only a few strategic locations, notably at the cylinder and piston, valves and guides, crankshaft bearings, and camshaft. When you think of engine wear, think in terms of thousandths of an inch. As an example, the cylinder bore diameter for one typical small engine is specified as 2.500–2.501 inches. The engine builder allows .001 inch manufacturing tolerance to account for expansion on hot days, contraction on cool days, minor wear in cylinder hones, and a host of other factors that influence the manufacturing process. But that's still pretty close. A human hair commonly measures about .003–.0035 of an inch. Split that hair into thirds and you gain a concept of the type of dimensions that are routine in engine building and repair.

The piston that runs in the cylinder bore has a specified diameter of 2.4950 to 2.4955 inches for its skirt, the lower part of the piston. That means at assembly, a large piston and a small bore would have a running clearance of .0045 inch, while a small piston and a large bore would have a running clearance of .0065 inch, equivalent to about two sheets of paper. In a water-cooled automotive engine, those clearances would be a little loose. Remember, though, we're dealing with air-cooled engines that operate over a much wider temperature range. Though an automobile engine practically never works at full load except when passing, you may put your small engine to work tilling a garden or driving

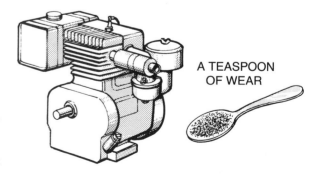

A TEASPOON OF WEAR

All of the wear particles from a badly worn engine probably wouldn't fill a teaspoon. This is why air-cleaner care and oil changes are so important.

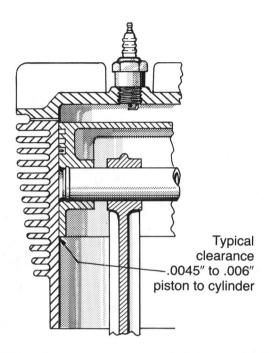

Running clearance between a typical piston and cylinder wall, above, is about the thickness of one or two sheets of paper. Even closer clearances are used between the connecting rod and crankpin bearing because there's less heat and less expansion here.

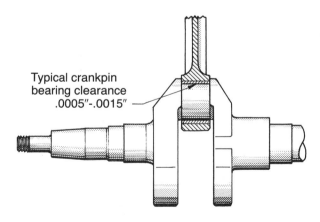

Typical crankpin bearing clearance .0005″-.0015″

a pump or generator and expect it to run at full load for hours on end. The clearances are needed for good oil flow and to compensate for the rapid expansion and contraction of the cylinder and piston.

Later, we'll talk about repair procedures on individual parts and how to actually measure wear. Very

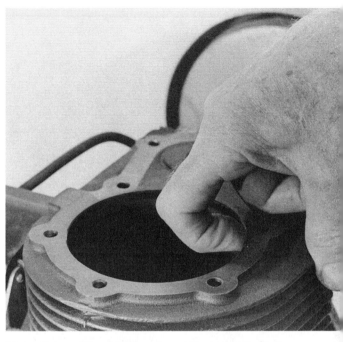

Check for a noticeable ring step at the top of the cylinder bore with your fingernail.

often, however, you'll want to make a quick evaluation as to whether your engine needs major rebuilding service. An experienced mechanic can often make a fair judgment quickly after removing the cylinder head. Here's how you can do it:

1. Unbolt and remove the cylinder head following the procedure on page 188. Turn the crankshaft so the piston is at the bottom of the stroke, then run your fingernail around the inside of the cylinder bore down in the polished area. If you feel a catch or roughness, use a strong light to inspect the bore for scores, scratches, or galling that indicate honing will be necessary.

2. Run your fingernail up towards the top of the piston travel where the piston stops and goes back down. Expect to find a ridge of carbon and combustion deposits at this point, but if gently scraping away the carbon at the sides—not the front and rear—of the cylinder reveals a perceptible ridge of metal, there's a good chance you have fairly severe cylinder wear. The thrust on the crankpin rocks the piston sideways slightly,

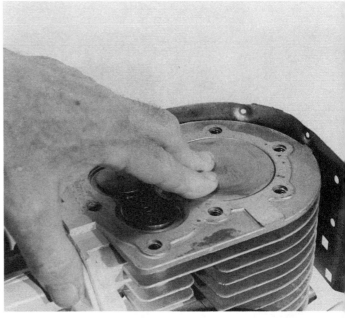

Feel carefully all around the cylinder bore. Use a good light to spot roughness, scoring, or rust.

Try pushing the piston back and forth at right angles to the crankshaft to check for looseness.

and most cylinder wear occurs at 90 degrees to the crankshaft. This is called a *ring step* or *wear ridge* and must be cut away even if all you do is install new piston rings.

3. Turn the crank so the piston is just below the ring step and use your fingers to push and pull the piston from side to side. Remember, we talked about a piston running clearance of .0065 inch or less. If the piston seems to slop back and forth more than that and feels loose, there is undoubtedly some cylinder and piston wear.

4. Place your fingers on top of the piston and gently rock the flywheel and crankshaft by hand. If you can produce and detect a significant movement of the crankshaft before the piston moves, or if you feel a slight rap or knock, the connecting rod bearing, the crankpin, the piston and piston pin, or all three are worn and loose.

5. Place your fingers against the end of the crankshaft or on the flywheel and tap the opposite end with a wooden block or a soft hammer. Now reverse the direction of the tap. You should feel very slight movement,

but it should not be over .004–.005 inch. Anything more than that indicates crankshaft thrust-bearing wear.

6. With the engine on a bench and the piston at top center, grasp the crankshaft or flywheel at first one end and then the other and lift up sharply as though trying to rock the engine on its base. If you feel a significant movement or hear a rap, the main bearings and probably the shaft are worn.

7. Turn the crankshaft so each valve, in turn, comes to the top of its travel. Grasp the valve head with your thumb and forefinger and try to shake it back and forth and from side to side. Expect to find some freedom but if the valves feel loose, the guides are probably worn. For now, disregard the condition of the valve seats and faces; they can be replaced or reground.

8. While checking the valve guides for looseness—with each valve at the top of its travel—measure from the top of the valve to the top of the cylinder block with a small steel square. The heights may not be exactly the same, but if one is significantly

To check for slop in the connecting rod and crankpin bearing, hold a hand on the piston and rock the crankshaft by turning the flywheel.

lower than the other it may point to a worn cam lobe. Remember that the exhaust valve may have a somewhat mushroomed head while the intake valve may have a flat head.

9. Many engines, especially those with aluminum cylinder blocks, have ring-shaped valve-seat inserts. Sometimes merely pushing sideways on the insert, with the valve open, will show perceptible looseness. Other times you'll have to place a drift against the insert and tap. Try squirting kerosene or light oil around the insert before tapping to detect movement. In many engines a loose insert can be pulled and replaced, but it's a shop job and the looseness suggests that the engine is probably worn out.

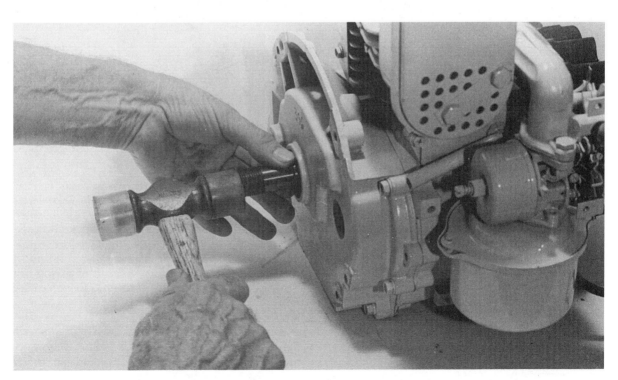

Too much crankshaft end play suggests internal wear. Try bumping the shaft first one way and then the other while feeling for significant movement. While not absolutely precise, this procedure is informative.

To check for loose main bearings, hold the engine down on a bench and try jerking the shaft up and down as shown above. The drawing below shows potential damage caused by an impact hard enough to bend a shaft—and by efforts to straighten it. A bent shaft is best replaced with a new one.

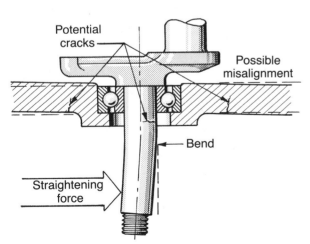

Although none of the preceding checks can tell the whole story without disassembly and actual measurement, they can certainly give you considerable insight into how badly your engine is worn. They can also guide you when buying used equipment. If the dealer doesn't want to let you remove the cylinder head for a look, offer to pay for a new head gasket and the labor of retorquing the head bolts. The small extra cost now may save you a lot later on.

Judging wear on two-stroke engines. The preceding checks are practical for four-strokes only, since most two-strokes do not have removable heads. They also use needle or ball bearings, which are not removable in the usual way. Unless twitching a two-stroke's flywheel back and forth produces discern-

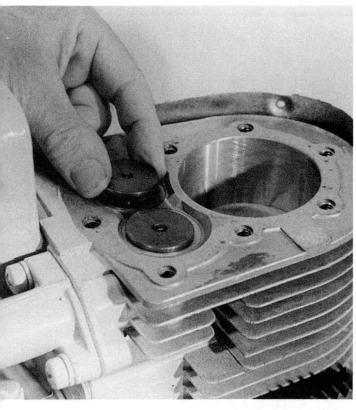

With the valve open, expect it to move sideways in the guide only a few thousandths of an inch.

Check for possible cam-lobe wear with a small square or depth gauge. One valve lifting significantly less than the other points to cam wear.

ible noise or movement, a more extensive check involving disassembly is required to accurately gauge the extent of wear that exists in the engine.

RUNNING PARTS VS. WORKING PARTS

Up to now, this book has dealt mainly with the engine's *running* parts such as the ignition system, fuel system, and governor. With the exception of a loose throttle shaft—which often indicates air-cleaner neglect or extremely dirty operating conditions—such parts bear little relationship to the engine's main, *working* parts and are easily repaired or replaced.

Servicing the working parts of your small engine

Tap valve-seat inserts lightly to see if they shift. Also place your fingers on them while tapping and try turning them.

If the throttle shaft is wobbly it's likely the engine has run in dirty conditions and may have inhaled a lot of dirt.

means restoring those parts to their original factory setting, or as close as possible. It also means you will almost certainly have to use a dealer's shop for resizing and other machine work. If, however, you are cautious and observant during disassembly, you should come through with flying colors. Remember, too, that even in a severely worn engine, the flywheel, cylinder head, and other parts may still be in good shape. That is the logic of short-block overhaul: *Use what's still good and replace what's worn.*

13

Cylinder Heads

Cylinder-head gasket leakage is one of the most common causes of hard starting. There are many reasons why a head gasket may start to leak. Chief among them is that the head and its gasket are subjected to the full impact of repeated combustion pressures and temperatures. While other working parts of an air-cooled engine also benefit greatly from the cooling action of the oil, the cylinder head must rely on outside air alone to conduct heat from the combustion chamber. Thus, anything that impedes the air flow through the fins on the head will also have a destructive effect on the head and gasket.

Many engines in lawn and garden service are subjected to large amounts of grass clippings and chaff.

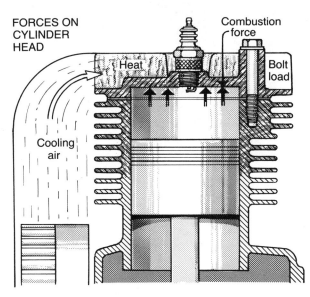

The cylinder head on a four-stroke engine must resist the full force of combustion loads—as well as expansion and contraction loads caused by extremes of heat—without warping or losing its gasket seal.

When some of those clippings become packed between the fins, trouble usually isn't far behind. Another source of what is basically overheating is an attempt to install an engine in a tightly enclosed hood or compartment. If you contemplate installing an engine in something of your own making, be sure to provide for free air flow to the cylinder head engine and block.

CYLINDER-HEAD TYPES

Most small gas engines are of the valve-in-block design, with the valve heads seated in the top of the block. The cylinder head on such engines is basically a flat plate, usually with a shallow inner cavity that forms the combustion chamber. On nearly all four-stroke engines the head comes off simply by removing the cylinder-head bolts. A few four-stroke engines have the valves and seats in the head. On those engines, removing the head is somewhat more complex and requires prior removal of the rocker-arm supports. You'll find the procedure in the following chapter on page 215.

On many small two-stroke engines, the head is cast as an integral part of the cylinder since there are no poppet valves and, thus, no need for valve access. The procedure for removing and inspecting a two-stroke's reed valves is also in the following chapter, on page 220.

CYLINDER-HEAD REMOVAL AND INSPECTION

You may decide to remove a cylinder head to replace a gasket, remove carbon buildup, service the valves, or perhaps inspect the cylinder bore. Removing the cylinder head from the typical valve-in-block—or *L-head*—engine is an easy job. Be careful,

Typical valve-in-block engine, above, has intake and exhaust valves beside the cylinder. Valve stems and springs are in the chamber beneath. Photo below shows inside of cylinder head that goes atop block. Note heavy carbon deposits in combustion chamber, which should be removed on reassembly.

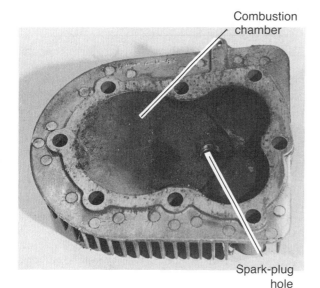

Combustion chamber

Spark-plug hole

however, to remember the location of each bolt as you remove it. The easiest way is to roughly trace around the head on a piece of stiff cardboard.

1. Begin by punching holes in the cardboard to match the location of each head bolt, then slip each bolt into its hole as you remove it.

2. If bolt lengths are all the same, you have no concern; if there are long and short ones, they must go back to the same locations on reassembly. Too long a bolt in the wrong place can damage cooling fins and cause a loose gasket, while too short a bolt can cause inadequate gasket clamping.

3. With all the bolts out, the head may just lift off. Or, try pulling the starter rope or bumping the electric starter (with the spark-plug lead disconnected) to pop it free by compression action. You can also bump the head lightly with a soft-face hammer or block of wood held against it.

4. Inspect the head carefully for broken fins, signs of burning along the edge of the combustion chamber, or damaged spark-plug threads.

5. Use a putty knife to remove any adhering gasket material. Then, place a piece of 240-grit abrasive paper on a perfectly flat surface and draw the inside-surface of the head flat across the abrasive while pressing down firmly. That will not only clean up the head but will also show any warpage and high and low spots.

6. Check the severity of warpage by placing the head on a flat surface without the abrasive paper and trying to slip a .0015- or .002-inch feeler gauge under it. Unless

Typical two-stroke engine has integral head and cylinder. Note spark-plug hole at top.

On most four-stroke, L-head engines, the cylinder head is easily removed for valve service.

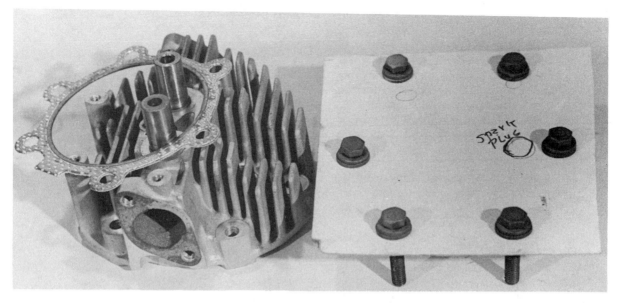

Unless you're absolutely certain that all head bolts are identical, make up a cardboard holder to maintain a record of their locations. This is a must particularly when bolt lengths vary between long and short.

The intake valve at left is warped, while the exhaust valve at right is badly burned. Both must be replaced.

warpage is severe, the procedure outlined in Step 5 should clean the head up and bring it back to flat. *A badly warped head must be replaced.*

Checking the valves. Even if you are just doing a head-gasket replacement, turn the crankshaft to bring each valve in turn to its maximum lift. Use a

small, bright flashlight to look under each valve and at its seat. If both valves and both seats look good proceed with the gasket installation. If they appear rough, burned, or cracked, a valve job is definitely in order. The procedure begins on page 201.

Checking and cleaning spark-plug threads. One of the points made in Chapter Five was that a spark plug can frequently become stuck in their threads, particularly if the head it screws into is aluminum. To avoid such damage, some manufacturers recommend an anti-seize thread lubricant. Be sure to use just a small amount on the plug threads only; lube introduced into the threads in the head can easily find its way onto the spark-plug electrodes.

Clean up the spark-plug threads *before* reinstalling a cylinder head by running a tap or tool made especially for the purpose all the way through the threads. That way, the cuttings won't go into the engine. *Do not install the head if the threads are damaged.* The compression and exhaust gases could cause the spark plug to blow out of the loose head threads with the force of a bullet, or—at the very least—the gases can escape past the threads, diminishing power and possibly causing severe head erosion and cracking.

In many cases the cheapest and most practical repair is a new cylinder head. The next choice is to take the head to a shop with the proper tools needed

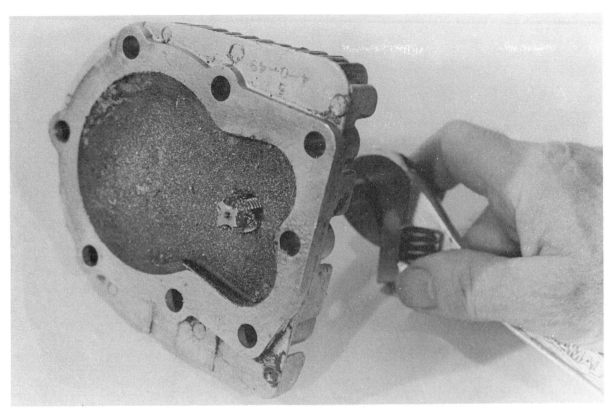

Whenever you have the head off, it's a good idea to run a spark-plug cleaning tool through the plug hole. A power tool top such as a table saw or drill press, plus

to cut the threads for a *helical-coil insert* and fit the spring-like device into the new threads. When completed, the helical coil provides very satisfactory threads for the plug.

If the engine is an antique for which no replacement head is available, have a machine shop bore out and thread the old plug opening and then fabricate a bushing-type brass or bronze insert with internal and external threads. The external thread should be a snug fit, actually a few thousandths oversize, with the threads in the head. Install the bushing by heating the head in hot oil and chilling the insert. Install a spark plug into the insert, then use the plug to quickly screw it into the head threads. It is also a good idea to pin or stake the insert to keep it from turning.

Cleaning the cylinder-block surface. The top of the cylinder block must also be clean and flat. Bring the piston to top-center with both valves closed, then

some medium-grit abrasive paper will show up warpage and blemishes on a cylinder head. A little scouring will usually flatten the surface nicely.

wire-brush and scrape away carbon and old gasket material. Also clean the top of the piston but do not scratch or gouge it.

Before plunking the new gasket onto the cylinder block, run a flat mill file over the block surface. Do this carefully but firmly enough to show up and remove any burrs, high spots around the bolt holes, or other blemishes that might interfere with good gasket seating.

Head-Gasket Installation

The head gasket itself is made of a crushable metallic material designed to conform to both head and block surfaces and form an airtight seal when the two are tightened together. Assuming the new gasket is the one specified for your engine make and model:

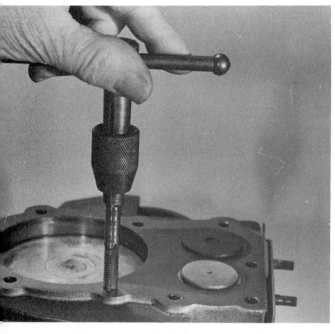

Clear head-bolt holes of rust and debris.

1. Before installing the head and gasket, be certain the bolt holes are clear of rust and debris. A tap should run in freely with light finger pressure.
2. Carefully place the gasket in position on the block and check that the bolt holes all line up perfectly.
3. Gently lay the head onto the new gasket, again making certain all the holes line up. Then start two bolts into their respective holes at opposite sides of the head and run them in with your fingers. *All bolts should first be turned in with fingers only.*

To remove hard carbon deposits, try a little engine cleaner followed by light wire brushing to finish off the piston and valve area.

Torquing Down The Head

Although you may not realize it, a bolt must actually be stretched to exert a clamping force. Moreover, that clamping force must be greater than any force that is trying to loosen the part retained. The head bolts on your small engine, then, must not only be stretched to a greater tension than the force of combustion, their clamping force must also compensate for gasket crush as well as expansion and contraction.

Place gasket in position with holes aligned.

Installing A Helical-Coil Insert

1. Torn or stripped head-threads are tapped for the helical coil with a special tool.

2. The coil insert is screwed onto an installing tool, then threaded into the head.

3. Once the coil is installed, the end-piece is broken off leaving new threads for the plug.

Turn in two opposing bolts with your fingers.

wrench. If something such as dirt packed in the bolt hole, rust, carbon, or damaged threads causes the bolt to turn hard, you may pull the specified torque without stretching the bolt enough to compress the gasket properly. To avoid this, clean out all head-bolt holes. Compressed air can help, but if the holes appear to have rust or other material in them, run in a tap of the proper size and clean up the threads. A small, round wire brush of the kind used for cleaning guns is also helpful. The end result should be clean, dry threads.

The same thing applies to the bolts themselves. If you have a power-driven wire brush on a bench grinder, go over each bolt until it is thoroughly clean. If you find stubborn deposits or bad threads, you may have to run a threading die over them. Discontinue the procedure if the die starts to bite metal; the threads on the bolt may have a slightly different profile than those on the die.

As a final check, run each bolt in with your fingers until it touches the head or its seating washer. If necessary, no more than a very light twist on the socket wrench should run them in freely. Investigate any bolts that seem to drag or be tight. An aluminum paste-lubricant is often recommended for bolts threaded into aluminum cylinder blocks; otherwise use any oil or grease recommended by the engine manufacturer.

That's where torque measurements enter the picture. Experiments have shown that a given bolt and thread diameter will approximate the desired tension when it is tightened with a given amount of twisting action by the wrench. In truth, however, all you're measuring is how hard you're pulling on the

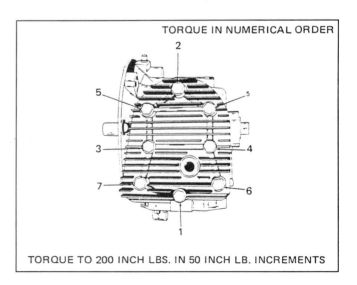

Head bolts are tightened and stretched so their tension is greater than forces trying to lift the head.

Tecumseh head-bolt tightening sequence, L-head engines less than 8 hp *except* VM and HM models.

Sometimes the head-bolt torque eases off after your engine has run for a while. Make a periodic check and retorque if necessary.

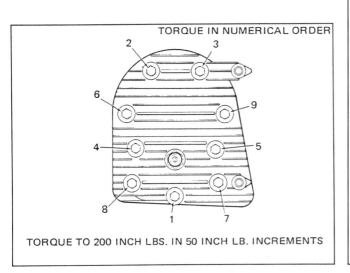

TORQUE IN NUMERICAL ORDER

TORQUE TO 200 INCH LBS. IN 50 INCH LB. INCREMENTS

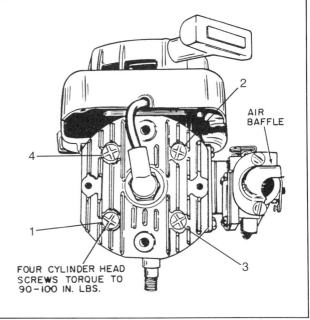

AIR BAFFLE

FOUR CYLINDER HEAD SCREWS TORQUE TO 90-100 IN. LBS.

Head-bolt tightening sequence, Tecumseh VM and HM80-100 model engines.

Head-bolt tightening sequence, for use on early Tecumseh two-stroke.

Tecumseh Torque Specifications
L-Head Engines 8-Hp And Larger

	INCH POUNDS	FT. LBS.
Cylinder Head Bolts	180 - 240	15 - 20
Connecting Rod Lock Nuts	86 - 110	7 - 9
Mounting Flange or Cylinder Cover	100 - 130	8.3 - 10.8
Flywheel Nut	600 - 660	50 - 55
Spark Plug	220 - 280	18.3 - 23.3
Carburetor to Cylinder	72 - 96	6 - 8
Air Cleaner to Elbow	15 - 25	1 - 2
Air Cleaner Bracket to Carburetor	15 - 25	1 - 2
Tank Bracket to Housing	35 - 50	2.8 - 4
Tank Bracket to Cylinder (5/16" Lower)	150 - 200	12.5 - 16.6
Tank Bracket to Head Bolt	150 - 200	12.5 - 16.6
Starter-Top Mount Recoil	40 - 60	3 - 5
Belt Guard to Blower Housing	25 - 35	2 - 3
Flywheel Screen & Pulley	72 - 96	6 - 8
Stationary Point Screw	15 - 20	1 - 1.2
Blower Housing Baffle to Cylinder	72 - 96	6 - 8
Blower Housing to Baffle or Cylinder	48 - 72	4 - 6
Breaker Point Cover	15 - 25	1 - 2
Magneto Stator Mounting	72 - 96	6 - 8
Breather to Cylinder	20 - 25	1.2 - 2
Oil Drain Plug 3/8 - 18	80 - 100	6.6 - 8
Blower Housing Extension to Cylinder	72 - 96	6 - 8
Stub Shaft Bolts to Flywheel	100 - 125	8 - 10
Motor Generator all Mounting Bolts	65 - 100	5.5 - 8

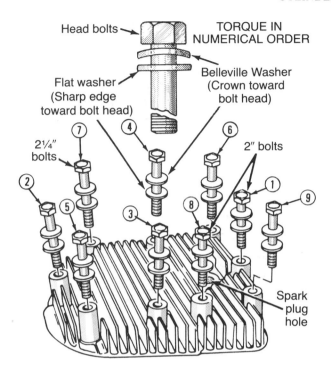

Head bolts

TORQUE IN NUMERICAL ORDER

Flat washer (Sharp edge toward bolt head)

Belleville Washer (Crown toward bolt head)

2¼" bolts

2" bolts

Spark plug hole

Head-bolt tightening sequence for use on Tecumseh L-head engines 8-hp and larger.

Using a torque wrench. From here on you'll be tightening the bolts step-by-step until they reach the desired stretch as measured by your torque wrench.

You'll also be compressing the head gasket so it more or less flows and distributes itself evenly between head and cylinder block.

This latter point makes it important that you tighten the bolts in the correct sequence and that you do it gradually, in stages. Although head-bolt torque specifications and tightening sequences are provided for popular Tecumseh and Briggs & Stratton engines, again, be sure you have the specs for your engine make and model. I can't recommend trying to get by without a tightening-sequence diagram, but if you must, the following technique should work out about right:

1. Pick two bolts as close as possible to opposite each other on opposite edges of the cylinder head, preferably lengthwise. Those will be numbers one and two.
2. Pick two more bolts, again at more or less opposite ends of the head, that are pretty much diagonal to each other. Call them three and four.
3. Repeat the diagonal cross-pattern on the other side of the head. Those bolts will be five and six.
4. If any bolts remain, they will be seven and eight.

The object is to work back and forth evenly. Tighten in small amounts until the bolt firms up a bit and then go to the next in sequence. Increments

Briggs & Stratton Head-Bolt Torque Specifications

BASIC MODEL SERIES	Inch Pounds	Meter Kilopond	Newton Meter
ALUMINUM CYLINDER			
6B, 60000, 8B, 80000, 82000, 92000, 94000, 110000, 100000, 130000	140	1.61	15.82
140000, 170000, 190000, 220000, 250000	165	1.90	18.65
CAST IRON CYLINDER	Inch Pounds	Meter Kilopond	Newton Meter
5, 6, N, 8, 9	140	1.61	15.82
14	165	1.90	18.65
19, 190000, 200000, 23, 230000, 240000, 300000, 320000	190	2.19	21.47

of about 50 inch-pounds are commonly recommended. As you approach the final torque, reduce this to 25 inch-pounds. When you go for the final tightening, bring the wrench onto the final torque value on the scale and hold it for a few seconds. You'll probably see the bolt creep slightly and have to add a touch more pulling force.

Even after tightening, the bolts may lose some of their initial tension after the engine runs. Run the engine after assembly until it is thoroughly warm, let it cool, and then go over the bolts with a torque wrench again. In my own experience I've found that after running the engine under load—perhaps a full season—a third torquing is good insurance against gasket failure.

COMPRESSION RELEASES

On some small two-stroke engines you may find a cover plate retained by two capscrews on top of the cylinder head. This is a compression release to ease starting. Normal service for this part is a thorough cleaning of the cavity under the plate, together with a check to see that the compression release ports are open and free of combustion deposits. The release valve may be a bimetallic part that responds to heat, or a reed valve that responds to pressure after the engine starts. Before removing the valve observe carefully which side is top and bottom and reinstall it the same way. If the valve shows damage from heat or corrosion, replace it. Seal the cover plate on assembly with a light coat of silicone gasket sealant on the mating surfaces.

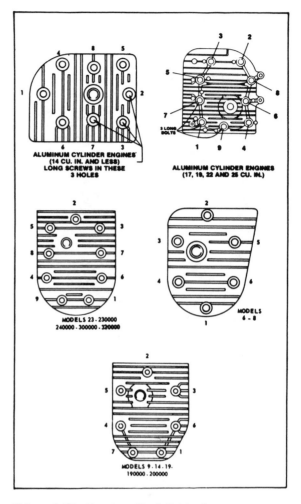

Briggs & Stratton head-bolt tightening sequences.

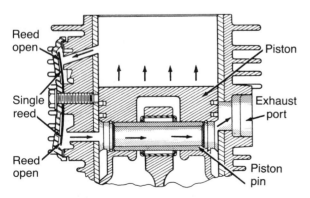

Compression release on a two-stroke engine. Part of the compression force passes the reed valve and goes

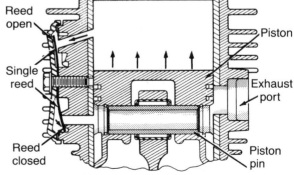

through the piston pin into the exhaust port. When the engine starts, lower part of the reed seals the passage.

14

Valves, Seats, and Guides

Good valve-seating is vital to maintaining compression in a four-stroke engine. If you visualize the valves snapping open and closed over a thousand times a minute at most working speeds, it's not hard to see why the contact faces become worn. The constant sliding of the valves in the guides also causes wear.

The intake valve will usually outlast the exhaust valve since it is exposed to a flow of cooling gasoline and air. The exhaust valve must cope with a constant flow of red-hot gases and may run at dull-red heat under load. Nevertheless, both valves must maintain an airtight seal on the compression and power strokes.

Fortunately, servicing the valves on your small engine is a common job that can be done without stripping the engine down to the bare cylinder, and—sometimes—without even removing the engine from the equipment it powers.

VALVES AND RELATED COMPONENTS

Most valves, especially exhaust valves, are made of at least two different materials welded together between the head and stem. The valve-head alloy is selected for high heat resistance and the stem alloy is selected for toughness and wear resistance. If the engine is expected to endure severe service and extended full-load use, still a third alloy of super-hard Stellite may be welded around the face of the seating area. The small additional cost of Stellite facing is usually worthwhile when buying a replacement exhaust valve.

Valve rotators. Small engines that run under almost constant load and speed conditions are likely to develop frequent exhaust-valve burning. Typically on farm and industrial equipment, the engine is started, the speed is set, and the load is almost steady hour after hour. With leaded fuel, the result is often a buildup of combustion residues and lead deposits that don't exit the engine because there is little expansion and contraction to break them loose.

When a piece of that residue does break off, it can lodge between the exhaust valve and seating surfaces and prevent the valve from fully closing. The result is that hot gases leak past the partially open valve, which begins to rapidly burn away around the edges. A valve rotator fitted at the lower end of the valve

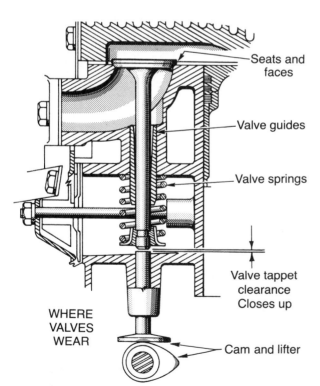

WHERE
VALVES
WEAR

Seats and faces

Valve guides

Valve springs

Valve tappet clearance
Closes up

Cam and lifter

Valves and seats wear gradually at a number of points.

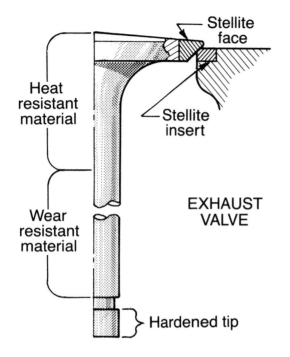

Most valves are made of at least two materials. The valve head is subjected to high combustion-chamber temperatures and is of a different alloy than the stem.

stem causes the valve to twist slightly each time it opens and closes. The twisting action tends to scour the seating faces and makes it harder for any combustion debris to become trapped between them.

Though the problem of combustion deposits may be reduced with the phasing-out of lead in gasoline, it can still occur if the engine uses excess oil and the oil additives build up similar deposits. Remember, too, that the deposits on the valve head and neck areas contain lead and lead-bromide compounds that can be extremely irritating to nose and eyes. They may also be toxic. Although a stiff wire brush in a bench grinder is a good way to remove the deposits, wear a respirator and goggles to avoid breathing the dust or getting it in your eyes.

Valve Seats

Some small engines, particularly older models, have both intake and exhaust valve seats simply cut and ground into the cast-iron cylinder block. Many modern engines will have an exhaust-seat insert of hardened metal, and aluminum-block engines will

have inserts in both intake and exhaust seats. Intake-seat inserts may be available as both original equipment and service parts for cast-iron engines, and on some larger, heavy-duty engines the exhaust inserts may be Stellite. Like other seat inserts, these are shaped like narrow rings and are driven, pressed, or shrunk into shallow counterbores surrounding the seat areas. Stellite valve seats resist the constant pounding of the valves seating under stiff spring pressure—particularly in the absence of lead in fuel, which formerly provided a cushioning effect between the valve and seat.

In most cases, both in-block seats and seat inserts can be cleaned up to a new precision seating face with the use of a seat-cutting tool or seat grinder. Sometimes, if the wear is very light, *lapping*—or rotating a new valve lightly coated with valve-grinding abrasive against the seat to match surfaces and clean up wear and pitting—will be sufficient.

Valve Guides

Earlier, it was suggested that one way to determine the extent of wear in an engine is to grip the head of the valve when open and rock it from side to side. This, of course, is only a rough check, but if the valve seems wobbly the guide is probably worn into a slightly bellmouthed shape.

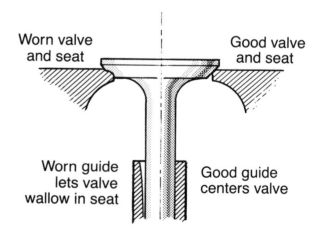

Good valve at right has a narrow seat area. Valves that have worn into a semi-rounded groove or show burning must be reground or replaced, and their seats must be resurfaced.

There are two reasons for correcting worn valve guides. First, if the valve wobbles in the guide it does not always seat in the center of the valve seat and will leak, causing rapid valve and seat wear. Equally important, if you are cutting or grinding the seat, the guide is used to center the pilot that guides the cutter. If the pilot wobbles in the guide, the seat will not be cut on true center and you'll be trying to seat a round valve on an off-round seating surface. Always check and repair valve guides as needed before proceeding with a valve-seat refacing job.

Valve Springs

All too often, the importance of properly functioning valve springs is overlooked when servicing the valves. These springs must snap the valves tightly closed at the end of each valve opening, sometimes over a thousand times each minute. This

means that the spring characteristics must be worked out very carefully by the engine builder if the valves are not to bounce, flutter, or pound into their seats.

Always mark or tag the intake- and exhaust-valve springs on disassembly. In many engines they are identical, but in many others they are distinctly different although this may not be apparent to the eye. The only way to be sure is to check the part numbers. Note also if the springs have damper coils. Damper coils can be spotted as a few turns of the spring at one end that are more closely spaced than most of the coils on the spring.

High-speed movies of valve and spring action run back in slow motion show the valves to be bouncing and the springs surging back and forth. The damper coils introduce a change in the spring harmonic characteristics at one end and thus reduce spring surging. When installing such springs, the damper coils usually go towards the non-moving end of the spring, which seats against the cylinder block or head.

VALVE-TRAIN OVERHAUL—L-HEAD ENGINES

This section deals with reconditioning or replacing valves, seats, and valve train components on a part-by-part basis. Once you've removed the cylinder head as shown on page 188, the valves, valve springs, and valve lifters are easily removed following the steps in the accompanying photos. Though a Briggs & Stratton engine is shown, the procedure is similar for other popular small gas engines of valve-in-block design.

Where To Begin

Because the valve guides are so essential to proper valve-face seating, as mentioned before, this is where all valve work should begin. It is also an area where information for your engine make and model is essential. In some engines, for instance, the guides are simply part of the engine-block casting. In others, they are pressed in and may be made from a number of different materials and require any of several techniques to service them.

One engine maker does not offer replacement guides but, instead, sells valves with oversize stems

Shown here is a valve spring with closely spaced damper coils at one end.

This lever-type tool is used to compress relatively light valve springs as shown.

to be fitted to guides that have been reamed oversize to accept them. Other engine builders have special reamers and/or puller tools to cut out or pull worn guides. In all cases, certain critical dimensions and procedures must be observed. Five minutes of note-taking with a manual open on the dealer's counter should give you what you need.

Even so, you may find that you need some special reamers and, as a beginning, you may need a special gauge to determine just how badly your valve guides are worn. All in all, this is one of those circumstances in small engine repair where, unless you have a well equipped shop and are familiar with such work, you're better off having the job done at an authorized repair shop.

Loose valve-seat inserts. Considering that the valve seats on most modern engines are actually valve-seat *inserts,* it is important to remember that they often loosen in service. Loose inserts are unable to conduct heat from the valve head to the block, and they overheat as a result. The insert then distorts and expands against the walls of the counterbore and loosens some more.

While loose inserts are seldom a serious problem with cast-iron blocks, they can be very troublesome with aluminum blocks. Since aluminum is softer

than the hardened seat insert, the aluminum can actually be pushed out by the expansion of the insert if the insert is overheated. Once good heat transfer to the block is lost the process progresses rapidly.

Therefore, before cutting or lapping seat inserts, be certain the insert is tight in the counterbore. As mentioned in Chapter 12, a loose insert can often be detected by trying to rotate or move it with your fingers, or by squirting kerosene around it and then tapping lightly with a small hammer.

Servicing Valves And Valve Seats

The simplest form of valve-and-seat resurfacing is *lapping,* as mentioned earlier. You'll need a valve-grinding compound sold for this purpose at automotive supply stores. A special valve-driving tool with a crank handle to rotate the valve alternately one direction and then the other is also handy if you plan on doing a lot of valve grinding. But for practical purposes, a simple tool consisting of a small rubber suction cup on a length of wooden dowel will suffice. These tools are also sold at automotive supply stores, but you can make one by borrowing a suitable cup from a child's toy. To use it:

This scissor-type tool is better, especially for stiffer springs. Valve spring and retainer are worked free as

a unit, and the valve is lifted out. The tool requires only one hand to operate, leaving the other hand free.

1. Begin by adhering the cup to the valve head.
2. Coat the valve-seating surface lightly with a fine valve-grinding abrasive, then use your palms or fingers to rotate the dowel back and forth and turn the valve face against its seat.
3. Rotate the valve through about 45 degrees, lift and turn it, and repeat the process. After doing this a short time, wipe away the compound and you should see a narrow, gray area where the abrasive has done its work between the two seating areas.

The pitfall in seating valves by lapping is that it's easy to wind up with a rounded, too-broad seating contact area. If this ring of contact is wider than about ³⁄₆₄ to ¹⁄₁₆ inch, it becomes a trap for small carbon particles that can hold the valve off its seat. The cure for that condition is called a *narrowing cutter.* Valve-seat angles will typically be either 45 degrees on both seats or 45 degrees on the exhaust and 30 degrees on the intake. Thus, if a cutter or stone shaped to a sharper angle is used down in the seat it will cut away some of the excess metal and leave a narrower surface.

Installing valve-seat inserts. One suggested

method of tightening a loose insert—as well as for securing a replacement insert—is to gently hammer or *peen* the metal around the top of the insert with a small flat end-punch.

Start with a light peening at three equally spaced points to center the insert and follow by peening all around to compress the metal of the cylinder block against the insert outer wall and flare a small flange over the top outer edge of the insert. Since this is far from an ideal repair, however, do not be surprised if the insert loosens again.

A better way is to have a machine shop make up an insert .010, .015, or .020 inch oversize on the outer diameter and to recut the counterbore to the same oversize—less .002 to .003 inch. Shrinking is a better installation method. If done properly, an insert will drop into place without driving or with only a light tap.

1. Begin by heating the cylinder block in boiling water.
2. Use your home freezer to chill and contract the insert. An even better method is to immerse the insert in dry ice and alcohol or gasoline until thoroughly chilled. *If you use gasoline, do so outdoors and well away from any source of flame.*

Final touch to a valve job is lapping the seat and face to a gas-tight seal.

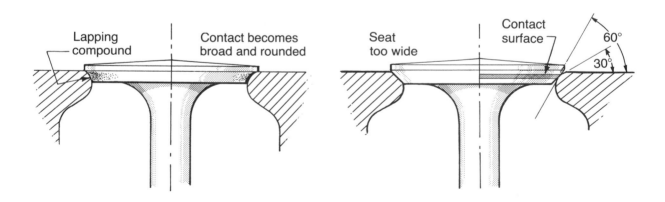

Extended lapping can cause a round seat and grooved valve, which trap combustion debris.

Removing substantial amounts of metal from a valve seat may make it too wide for good sealing.

3. Do not touch the hot block or chilled insert with your bare hands; hook the insert with a piece of bent wire and quickly drop it into the counterbore in the block.

4. Use a valve or like tool to push the insert firmly against the bottom of the counterbore. Work quickly, because the insert will lock in place almost instantly.

Removing inserts. Tecumseh does not recommend insert removal. If inserts are not provided with the needed configuration to get under them with a puller, they are very difficult to remove without damaging the counterbores in which they seat. It is important that they be lifted or drawn out straight up rather than cocked in the counterbore.

One tool that does this neatly consists of a cup-shaped puller body with a threaded puller screw that engages a washer-like nut. The nut has flattened sides so it may be canted and placed under the lower, inner edges of the insert and then pulled up flush against the insert to withdraw it. The nut portion of the tool is available in a number of sizes for different size inserts. The tool can also be easily made in the home shop.

Valve-seat grinding. After valve-guide service and seat-insert replacement, the center of the valve guide and the center of the insert have been disturbed and misaligned and must be brought back on a common axis. Even if neither was disturbed, the valve guide is used to center the cutting tool so the seat will be on center with the valve when it rides in the guide.

As mentioned, sometimes simply hand-lapping the valve to the seat will suffice, and it may be that using a new valve with clean, flat seating surfaces will do the trick. It is also true that a moderately worn or slightly burned valve may often be refaced and used again. When you clean up a valve face, however, it is usually necessary to remove so much metal that the edge of the valve becomes marginally thin. Some shop manuals show that this edge, or margin, is acceptable if it is $\frac{1}{32}$ inch thick and unacceptable if it is $\frac{1}{64}$ inch thick. Also, unless you are equipped to true up the valve face yourself, taking it to a shop will cost about as much as a new valve—which probably makes more sense.

Truing up the valve seat or insert can be done using either a cutter tool with carbide blades or by grinding with a stone contoured to the exact 30- or 45-degree angle. Stellite inserts will almost certainly require grinding in a suitably equipped shop.

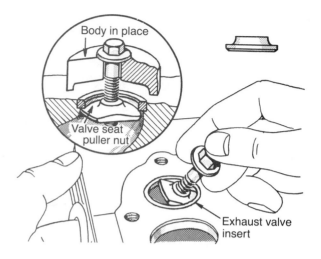

The solid body on this insert puller has a slot so the puller plate can be slipped into place and the body slid over the bolt.

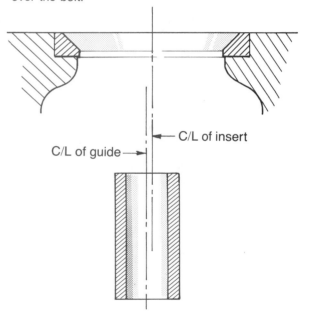

The valve guides should always be restored first and the guide bores used to center the seat cutter or stone.

A good mechanic will try several pilots in the valve guide to find one that fits snugly. Pilots vary slightly to compensate for worn guides. Such pilots are hardened rod-like members precisely ground to

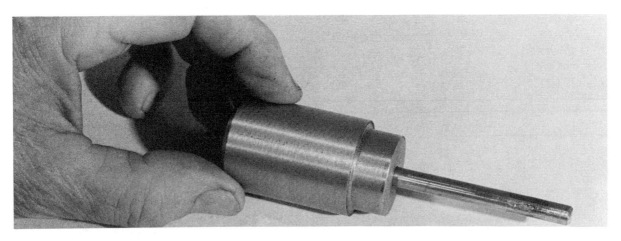

Seat-driver tool above has a pilot made from an old valve stem and pressed into place. Shown in use below, pilot keeps the insert square with valve guide as you tap the insert into the seat.

a slight taper so they will seat tightly in the guide bore. The pilot extends up past the seating surface and establishes a true center for the stone or cutter.

If an abrasive stone is used, it is important that the stone be trued up with a diamond tool first. Stones tend to wear in a groove at the narrow contact surface and will cut a rounded rather than flat surface if not trued up occasionally. Note that some engines specify a 46-degree angle for the valve seats. That is called an *interference angle* and is claimed to improve seating against a 45-degree valve face, though it isn't critical. A Stellite seat will require a stone with different abrasive characteristics than a standard seat.

The stone is rotated rapidly by a driving tool resembling an electric drill. The trick is to use a light touch and lift the stone after a few seconds of contact. If the engine is only partially disassembled, provide protection for the piston and cylinder bore so bits of grinding swarf do not get down between the piston and cylinder wall. One way is to smear the

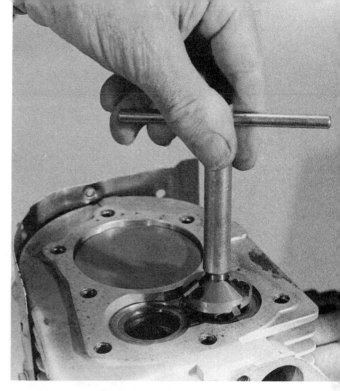

Carbide tool for cutting valve seats.

1. Loose valve seat can be turned or moved up or down. Check with feeler gage here.

2. Use center punch to tighten insert at three points, equally spaced.

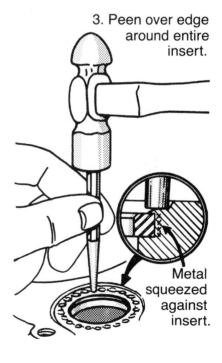

3. Peen over edge around entire insert.

Metal squeezed against insert.

To tighten a new insert, peen it with a flat-end punch, first at three equally spaced points to center it and then all around to secure it in the counterbore. Work carefully and slowly to be sure insert seats properly.

A light coating of Prussian Blue on the valve seat shows where the valve is contacting.

top of the cylinder block with grease and press a piece of heavy wrapping paper into it to cover the open cylinder.

If the initial cutting shows that the stone is producing too broad a seat, you can either switch to a narrowing stone now or continue until you've developed a clean seat and follow with a narrowing stone.

Locating the seat. A second factor in a good valve job, in addition to controlling the width of the seat, is locating the seat properly in relation to the upper valve-edge. The contact should not be up close to the edge or margin or down near the bottom of the seat face. Check seat location by smearing a thin coat of Prussian Blue on the insert face and dropping the valve into place. The above conditions happen most often on a seat that has been ground several times. In some instances you'll need a second, shallow-angled narrowing stone to clean away part of the insert at the top. The rule is that the finished contact area should be about 3/64 to 1/16 inch wide and

located about midway on the valve-seating face. One point to bear in mind is that you will generally be working with much smaller valves and seats than are common in auto and industrial engines. It takes very little pressure and very little cutting to do the job. If you take your engine to a shop accustomed to automotive work, they may be a bit heavy-handed.

An equally good job can be done on most valve seats with a hand-turned cutting tool. Again, a pilot is used, but rather than a high-speed grinding stone, a cutter with a number of carbide blades is rotated slowly by hand to shave the seat clean. Although not inexpensive, these seat cutters will adapt to many engines and will last a lifetime. The cutter blades are replaceable.

Finish-lapping. After cleaning up the valve seat it is good practice to lap the valve and seat lightly for perfect contact. Use a fine valve-grinding abrasive and coat the valve lightly with it.

As a final check, many mechanics will make soft

A valve that's only slightly worn can be cleaned up by turning in a carbide cutter such as this. Take light cuts and remove no more than necessary.

A better check, of course, is to place the old springs beside new ones if such are available.

Also, as mentioned in Chapter Four, valve springs do tend to corrode or rust. Always clean off the valve springs thoroughly and look at them in good light, preferably under a magnifying glass. If you see small pitted areas, corrosion, or rust, there is a good chance the spring will break in service.

Valve-spring retainers. In Chapter Three, it was mentioned that a valve-spring compressor can be extremely useful. That is because removing and installing the spring retainers can be frustrating when you're working against stiff valve springs.

There are several ways to retain the springs under partial compression on the valves. The simplest is a short pin passed through a drilled hole in the valve stem. Another fairly simple method—from the

pencil marks at several points around the seat. After dropping in the valve and rotating it, each mark should show contact all around the seat. The same check can be made with a very light coat of Prussian Blue on the valve face. There should be no point showing undisturbed bluing. If there are such spots, the cutter or stone may have wobbled on the pilot, you may have applied sideways pressure in the cutting process, or the valve is out of round. The latter case is likely only with a used valve.

Inspecting Valve Springs

Before reinstalling used valve springs, check them for squareness by seating them against a flat surface and then placing a steel square against them at several points. There should be no signs of cocking or off-squareness. If there are such indications, replace the springs.

Sometimes springs are fatigued and unable to provide the proper valve-closing action. One check for that is to place the exhaust and intake springs side by side and look for a difference in height. A short spring indicates that the spring has lost its elasticity.

Check old valve springs for squareness on any flat surface as shown.

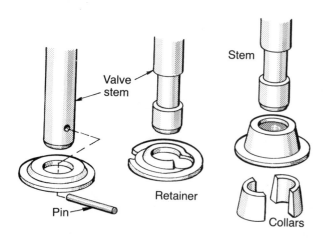

Three common valve retainers include a pin through the valve stem, a slotted washer that engages a notch in the valve stem, and tapered keepers that wedge into a matching taper in the valve-spring lower washer.

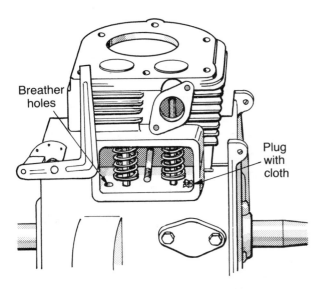

Breather holes in the valve-spring chamber are hard to see, yet valve keepers somehow manage to find and fall through them. Plug these holes with a scrap of cloth to avoid trouble.

standpoint of removal and replacement—is a dish-shaped washer with a hole that fits around the valve stem in a reduced section or notch, as well as a slot to one side that allows the washer to be pushed aside and dropped off or installed on the valve.

A third common retaining method follows automotive practice and uses two tapered collars, often called *keepers*. These fit around a reduced section of the valve stem and are held in place by a mating tapered bore in the valve spring retaining washer. Such valve-spring retainers are perhaps the most difficult to replace, since one keeper must be put in place and held there while the second one is placed on the opposite side of the valve stem and also held in place as the retaining washer is eased down over them both.

It must be admitted that removing and replacing the valve-spring retainers goes much easier if you're working with clean, dry parts, under a bright light on a bench. Even if you intend to leave the engine mounted on its machine, try to start off with dry parts and a dry valve chamber. You can oil the valve stems and guides later, but for now, keep them dry.

Valve-Train Reassembly

As any experienced mechanic can tell you, it's easy to drop a small retainer pin and particularly the small tapered keepers. Unless the engine is opened up so you can readily retrieve a dropped keeper, make a habit of feeling around the valve chamber for any holes or passages often provided as breathers or oil drains. Plug any such openings with a scrap of clean cloth so that a dropped keeper cannot find its way down into the crankcase. The latter misfortune usually calls for at least partial disassembly to get the keeper back. It's also a good idea to spread a catch cloth or towel under the work area so a dropped keeper doesn't get lost down in the tractor or other structure.

Before going ahead, make a final check to be sure you aren't putting an exhaust valve in the intake guide or vice versa. Also recheck the springs for their assigned valve. The next step requires compressing the valve spring, usually almost solid, against the top of the valve chamber while holding the valve down against the seat. The retaining washer must now be in place under the spring and ready for the locking device to be installed.

If you are limited to a screwdriver as a compressor tool, proceed as follows:

1. Pry up under the retaining washer until the spring feels solid.

2. Because a screwdriver tends to lift the spring in a cocked fashion and make installing the keeper difficult, find a strip of metal that can be wedged under one side of the spring as you pry it up. Properly done, you can now lift the other side with the screwdriver and hold the spring squarely compressed with one hand and have the other hand free to install the keepers.

3. If the keeper is just a pin, it isn't too difficult to slip through the hole in the valve stem with needlenose pliers. With slotted-washer-type keepers, you should now be able to slide the washer over to center on the valve stem. Ease down the screwdriver and remove the metal-strip prop from the other side of the spring, and your valve is now installed.

4. Tapered keepers can be installed the same way, except that you must hold one in place while positioning the other one. The standard method is to daub thick, sticky grease on the reduced area of the valve stem where the keepers seat.

5. With a daub of grease on the screwdriver tip—or by magnetizing the screwdriver—you can delicately maneuver each keeper into place so it is snug in the valve-stem recess as you lower the retaining washer over them.

All of the above goes much easier with a valve-spring compresser, which lifts the spring evenly and holds it there while you have both hands free to work. Also, when the keepers appear to be in place, you can ease the spring down slightly and make sure the keepers and washer are aligned during the lowering process.

With both valves in place, you can now remove the protective plugs from the valve-chamber holes and rotate the engine to open each valve in turn. Squirt some light engine oil down the valve stems for initial lubrication, and rotate the crankshaft a few turns to make certain the valve keepers are securely in place and that valve opening and closing is functioning normally.

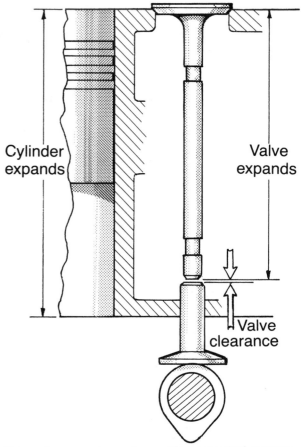

Valve clearance in engines with mechanical tappets compensates for the constant expansion and contraction of valves, cylinder, and other engine parts.

Adjusting Valve Clearance

Until the development of hydraulic valve lifters, all poppet valves operated with a specified amount of clearance or valve *lash* between the lower end of the valve stem and the lifter or tappet.

If you're unfamiliar with engine terms, a lifter, a tappet, and a cam follower are basically one and the same. The valve lifter rides on the cam profile and is lifted and lowered as the camshaft rotates at half engine speed. Remember, for every four strokes of the piston in a four-cycle engine the crankshaft turns twice; but there is only one intake valve opening and

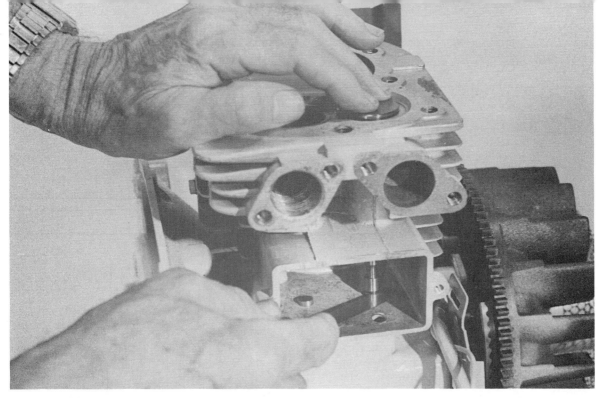

With the valve held down firmly and the lifter on the low point of the cam, this feeler gauge will not enter and metal must be removed.

Rig a guide on the grinder support to square the valve. Use a light touch to avoid taking off too much.

After grinding, a feeler of the specified thickness should just slip in and out between lifter and valve with a slight drag.

one exhaust valve opening in those two crankshaft revolutions. Therefore, the camshaft turns at half engine speed—or *once* for every four piston strokes.

With the mechanical lifters typical to small engines, valve clearance is necessary because of the constantly changing conditions in an operating engine. Picture the situation if the valves were adjusted so the stems were just barely contacting the lifters and the valve faces were just barely on the seats with the engine cold. As the engine started, the valves would soon expand a few thousandths and the valves would be unable to seat since they would now be held off the seat by the lifters. At the same time, the cylinder block and every other working part of the engine would be warming up, expanding, and changing dimension slightly. As the load on the engine changes, this expansion and contraction goes on constantly. Moreover, as the valves wear into their

seats slightly, the perfect adjustments would again be upset and the valves would be held off their seats.

The preceding problems are solved by providing a small clearance between the lifters and the bottoms of the valve stems. The exhaust-valve clearance is always greater because it runs hotter and expands more. Though the clearances will vary according to heat and loading as the engine operates, they will ensure that the valves always seat and will allow for a bit of face and seat wear.

With small air-cooled engines valve clearance is routinely set with the engine cold. In many engines it can only be done by grinding a little metal off the bottom of the valve stems until the correct clearance is arrived at. To make that adjustment:

1. Turn the crankshaft until both valves are closed and the piston is at top center on the compression stroke. Be sure the piston is in

Piston top

Intake valve

Valve spring

Push rod

Rocker arm

Workings of a typical overhead-valve engine. This one is a late-model Tecumseh.

this position; otherwise, the lifters will be partially on the cam lifter faces, or *lobes*.

2. Try inserting a feeler gauge between the lifter and the end of the valve stem while holding the valve head down firmly with your fingers.

3. In most cases—especially with a new valve—it will be impossible to enter the feeler gauge. Pressing down on the valve head and rotating it will show that the valve turns easily because the lifter is holding it up off its seat.

From here on, it's cut and try. The best way to grind off the necessary metal is with a commercial valve grinder since it assures that the end of the stem will be flat and not angled. Lacking such a machine, the job can be done on a conventional bench grinder. The main things to avoid are removing too much metal and grinding the end of the valve stem so it's angled or even slightly cone shaped. One way to avoid the latter is to rig a small V-block on the grinder tool support so the stem can be supported against the side of the stone and turned slowly.

Don't try to take off too much at one time. After

the stem is trimmed enough so the valve now seats, proceed in very small amounts and recheck with a feeler gauge each time. It doesn't take long to learn to estimate how much you're removing in terms of thousandths of an inch with each touch of the grinding wheel. The final clearance check—made with firm finger pressure on the valve head—should be felt as a light but definite drag on the gauge, which should enter the clearance gap with a light push. Don't panic if you go a thousandth or so over, but avoid excessive clearance since it will cause the valve to seat too hard and may result in the valve head eventually snapping off.

Adjustable lifters. A somewhat more sophisticated method of adjusting valve clearance is used on some engines, particularly those intended for heavy-duty industrial service. Here, a finely threaded flat-head capscrew is threaded into the upper end of the lifter. The screw has snug threads and is further locked against turning by a check nut. The lifter has flats on it so it may be held from turning while adjusting the screw.

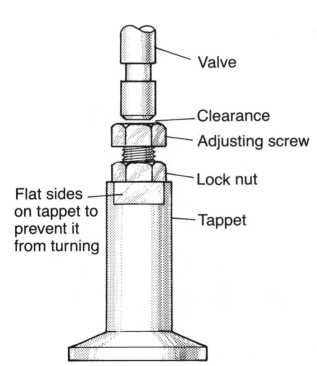

Screw-type valve adjustment usually requires two and sometimes three wrenches to loosen the lock nut, turn the adjusting screw, and hold the tappet.

To adjust valve clearance in such engines, you'll need two thin-end wrenches.

1. With one wrench held on the lifter, use the other wrench to first loosen the locknut and then turn the adjusting screw up or down.
2. Check with a feeler gauge as previously described. The valve spring will hold the valve on its seat.
3. When the adjustment produces a slight drag on the feeler gauge, tighten the locknut firmly and then recheck the clearance to be sure the adjusting screw didn't move.

VALVE-TRAIN OVERHAUL—OHV ENGINES

Though a valve-in-head—or *overhead valve*—design is the norm in automobile and aircraft engines, it has not been widely used in small gas engines. One reason may be the additional cost of the necessary push rods, rocker arms, and rocker-arm supports. Overhead valves also give an engine a higher profile, which may not be convenient in some installations. A third factor is the need to deliver oil to the rocker arms and valves and then drain it back to the crankcase. Since some oil will always accumulate in the rocker-arm housing, seals are needed between the rocker housing and the cylinder head and at the tops of the push rods.

Overhead-Valve Disassembly

The principle overhead-valve small engines made in America require very little modification of their valve-servicing technique from that used in L-head engines. Some specific disassembly and assembly procedures must be followed.

To remove the cylinder head, you must first remove the rocker-arm housing that mounts on top of it. That allows access to one of the cylinder-head bolts beneath the housing. Before removing the housing, however, you must remove the valve springs, which are retained in the conventional manner and prevent the housing from being lifted free.

1. To remove the valve springs, begin by loosening the locknut atop the rocker arm.
2. Run the lower, adjusting nut up into the

Overhead-Valve Disassembly

1. Run lower, adjusting nut up into the rocker arm, then remove snap-ring retainer at end of rocker shaft.

2. With valves supported inside the cylinder to prevent them from dropping, compress springs as shown. Then remove valve-spring keepers by levering the special tool against the rocker shaft.

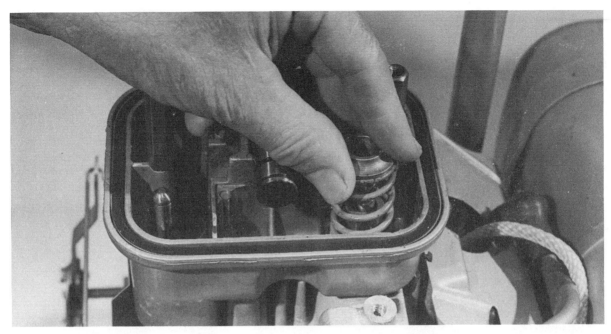

3. Once tapered keepers are removed, springs and washers are easily lifted out.

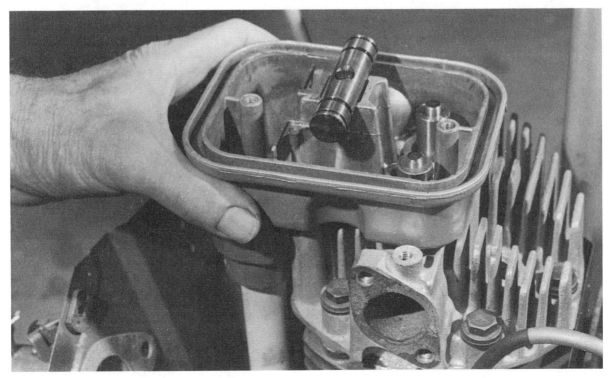

4. Rocker housing is now ready to be lifted off.

If your two-stroke seems a bit anemic, don't be surprised to find clippings or other debris under a reed.

rocker arm, then remove the snap-ring retainer from the end of the rocker shaft and slide the rocker arm off.

3. Take out the spark plug and hand-turn the engine until the piston is partway down. Then thread in a length of sash cord between the piston top and the valves to keep the valves from falling when you compress the springs.

4. Compress the valve springs with a tool that levers up against the rocker shaft, remove the valve-spring retainers, and lift out the springs.

5. With the rocker arms and valve springs removed, take out the housing retaining bolts and lift housing and rocker shaft free.

Replacing valve guides. On overhead valve engines with aluminum heads, the valve guides are shrunk into the cylinder head. Removing the guides requires heating the cylinder head in crankcase oil at 375 to 400°F. Use any handy container large enough to hold the oil and a wire rack or other means to hold the head up off the bottom.

1. Heat the container of oil on a hot plate until it smokes.

Tecumseh Overhead-Valve Engines

TORQUE SPECIFICATIONS		
	INCH POUNDS	FT. LBS.
Cylinder Head Bolts	180 - 240	15 - 20
Connecting Rod Screw or Nut	86 - 110	7.2 - 9.2
Mounting Flange Cylinder Cover Screw	100 - 130	8.3 - 10.8
Cylinder Cover - Flywheel End	120 - 160	10.0 - 13.3
Flywheel Nut	600 - 660	50 - 55
P.T.O. Shaft to Flywheel	100 - 125	8.3 - 10.4
Spark Plug	220 - 280	18.3 - 23.3
Intake Pipe to Cylinder	72 - 96	6.0 - 8.0
Carburetor to Intake Pipe	48 - 72	4 - 6
Air Cleaner Elbow to Carburetor	25 - 35	2.0 - 2.9
Air Cleaner Bracket to Elbow	15 - 25	1.3 - 2.1
Air Cleaner Bracket to Carburetor	15 - 25	1.3 - 2.1
Tank Bracket to Cylinder (5/16" Lower)	150 - 200	12.5 - 16.6
Tank Bracket to Cylinder (1/4")	96 - 120	8 - 10
Starter-Recoil-Top Mount	40 - 60	3.3 - 5.0
Starter-Electric-Straight Drive	140 - 170	11.6 - 14.1
Rope Start Pulley Mounting Screws	60 - 75	5.4 - 6.8
Rocker Arm Box to Cylinder Head	80 - 90	6.6 - 7.5
Rocker Arm Cover and Breather (Screw to Stud)	15 - 20	1.2 - 1.6
Rocker Box Cover to Mtg. Stud	20 - 30	1.6 - 2.5
Magneto Stator Mounting Screw	72 - 96	6 - 8
Blower Housing to Baffle or Cylinder	48 - 72	4 - 6
Housing Baffle to Cylinder	72 - 96	6.0 - 8.0
Blower Housing Extension to Cylinder	72 - 96	6.0 - 8.0
Governor Rod Clamp to Rod (Screw and Nut)	15 - 25	1.3 - 2.1
Governor Rod Clamp to Lever (Screw and Nut)	15 - 20	1.3 - 1.7
Ground Wire to Terminal	5 - 10	.4 - .8
Toggle Stop Switch Nut	10 - 15	.8 - 1.4
Oil Drain Plug 3/8 - 18	80 - 100	6.6 - 8.3
Oil Drain Cap	35 - 50	2.9 - 4.1
Oil Fill Plug (Large Diameter)	Hand Tighten	

2. Meanwhile, arrange to support the head on the base of an arbor press so there will be space beneath to allow the valve guides to be pressed out. A mandrel or arbor punch that pilots in the guide bores and also clears the guide bores in the cylinder head will also be needed.

3. When the head is thoroughly heated, quickly but carefully invert the head on the arbor-press supports and force each guide out. *Again—do not attempt to handle the heated head with your bare hands.* Use the wire rack or other means to pick up and position the head for service.

4. After chilling the replacement guides in the freezer or in dry ice and alcohol, reheat the head in hot oil and press in the new guides from the top of the head.

Valve adjustment. The procedure for overhead-valve engines is similar to that for L-head engines with adjustable lifters, as shown on page 215. The difference is that adjustment is made either at the rocker arm or by turning the adjusting screw at the top of the push rod.

REED-VALVE SERVICE—TWO-CYCLE ENGINES

The reed valves serve the same purpose in two-stroke engines—admission of the air/fuel intake charge—as the poppet type intake valve in a four-stroke. Note, however, that there is no exhaust reed valve since the exhaust port is opened when the piston slides past it on the downward stroke. Reed valves are also used as automatic compression releases on some two-stroke engines and as breather valves on four-stroke engines. The compression-release feature reduces the compression for easier cranking and then drops out of action as the engine starts.

The basic point to understand about reed valves is that they do not have any mechanical actuation. There are no cams, lifters, springs, or push rods. The only force to open and close them is air pressure.

During the upward, intake stroke of the piston, the negative pressure inside the crankcase is overcome by positive atmospheric pressure and the thin, spring-like valves or reeds lift off the seat enough to admit the fuel/air charge. On the downstroke of the piston, the mixture in the crankcase is compressed and overcomes atmospheric pressure to press the reeds tightly against the reed plate and, thus, effectively seal the passage.

Inspection And Replacement

While poppet valves may look very much alike in all engines, reed valves may vary greatly in shape and in the way they are mounted. Inspection and servicing techniques, however, remain the same. As a starting point, the reeds and the plate on which they seat must be clean. Even small amounts of dirt, which would cause little concern with a poppet valve under heavy spring pressure, will prevent a reed from seating and seriously impair engine performance. That means the air-cleaner must seal perfectly to keep out the typical grass cuttings or other debris normal to small-engine operation.

Second, the reeds must be flat and undistorted, and the flat plate on which they seat must also be flat and not corroded or worn. In some engines the reeds may have a very slight reverse curvature to spring them lightly against the reed plate, but more commonly the reeds are flat. If inspection shows them curled away from the seating surface, they are suspect and should probably be replaced. One engine manufacturer gives an allowable tolerance of .010 inch for the reeds to be bent off the seat.

When installing reeds, be sure the proper side of the reed goes against the seat. That is because the reeds are stamped from a sheet of springy sheet steel and the stamping leaves a slight roughness or burr on one edge. In most cases replacement reeds will be marked in some manner to show which is the seating side, but if in doubt, always install them with the rough edge away from the seat.

Compression-release reeds are located on the side of the cylinder, rather than between the carburetor and crankcase, but the same general principles apply to their service.

15

The Crankcase

The crankcase, particularly in four-stroke engines, is usually cast integrally with the cylinder. Small two-strokes often have a separate crankcase; often, such as with chainsaws, these are highly specialized parts that actually form much of the working structure for the saw or tool. Such crankcases may also be formed of two halves, split in either the plane of the crank-shaft or at 90 degrees to it.

Remember, although the details and appearances of crankcases differ, the basic job they do remains the same and the service and repair requirements have much in common.

WHAT THE CRANKCASE DOES

Above all, the crankcase must provide a rigid, solid, and perfectly aligned support for the crank-shaft bearings. The crankshaft main bearings must also be perfectly aligned on a common center line.

In most small-engines, the basic crankcase and cylinder block includes an end or *cover* plate.

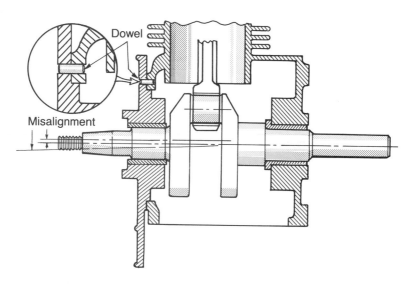

Locating dowels, above and below, establish a permanent alignment between crankcase and cover-plate. Without them, the necessary clearance in the cover plate bolt holes would allow slight shifting.

When your engine was built, the end or *cover* plate that retains one main bearing was installed and then doweled so it could not shift, and the crankshaft bearing bores were machined on a common axis in a single operation by passing a tool straight through from one side to the other. Similarly, the crankcase halves of small two-stroke engines are also assembled and then bored for bearings. They become matched pairs ready to receive precision ball or roller bearings.

Misalignment. It is still possible for misalignment to creep into the picture later when the engine is put to work. The constant pull of a belt or chain can cause a bearing to wear towards one side. We are now trying to run the shaft in two bearings that are angled or misaligned.

Crankcase loads. Another job of the crankcase is mounting the engine to the driven machine. In addition to side pulls from belts, there are almost always torque or twisting forces that work to distort the bearing lineup. Often, the simple drawing-down of the mounting bolts can induce severe structural loads on the crankcase if a tractor frame or other mounting structure is not flat and true or becomes distorted during use. Bearing trouble can eventually appear and its source may be quite mysterious. Unfortunately, it is common for power-equipment builders to use the engine as a structural frame part or stiffener for the machine.

Crankcase pressures. In Chapter Two, the importance of good sealing in crankcase-charged, two-stroke engines was emphasized. Without good crankcase seals and crankcase vacuum, the engine simply will not run. Thus, loose and leaking crankshaft bearings and seals on two-stroke engines must be replaced.

A common test of the overall health of a four-stroke engine is to measure crankcase vacuum. For a quick spot-check, if the engine has an extended oil-filler tube place the fleshy base of your thumb over the filler opening with the engine running at full throttle. You should detect a light but definite suction. If you have some form of automotive vacuum gauge, arrange a tight cork-and-tube connection in the oil-filler opening and read the vacuum directly. Expect to see the needle come off the peg and hold between 1/2 and 1 inch of mercury in a healthy engine.

There are two reasons for maintaining a vacuum or reduced pressure in the crankcase of a four-stroke

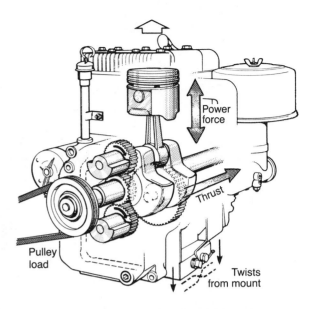

Combustion loads on main bearings and crankcase are considerable, as shown above. Heavy twisting forces also come into play (below).

engine. First, the combustion process introduces combustion gases, water, and sometimes raw gasoline into the crankcase. These contaminate the oil and can contribute to acid and corrosion. Since the oil normally runs above 212°F, those combustion by-products tend to remain in vapor form and can be drawn out of the crankcase by negative pressure through the breather system. The system works much like the positive-crankcase-ventilation or *PCV* system on your car. The waste products are normally discharged to the carburetor intake where

A warmed up engine in good condition will show a crankcase-vacuum reading of one-half to 1 inch of mercury as seen on this vacuum gauge. If not, suspect a clogged breather, bad shaft seals, or worn valves and rings. This, incidentally, is one way to gauge engine condition without reassembly.

they are taken in and burned or passed through the combustion chamber to the exhaust.

The second reason for crankcase vacuum is to prevent or reduce oil leaks. When the piston comes down, it compresses the crankcase vapors and without an escape passage the compression would force oil out around the crankshaft bearings. With negative pressure the oil is induced to stay in the crankcase. Some leakage of the seals will introduce fresh air past the crankcase seals. If the crankcase seals are worn too badly, the engine might also ingest dirt—most of which will be trapped at the lip of the crankcase seal, where it will act as an abrasive on the seal area of the shaft and increase wear on the seals.

Conditions other than bad crankshaft seals may cause a low crankcase vacuum, and that's why the vacuum check is important. Excess blow-by from bad piston rings, cylinder and piston wear, leaking valves, a sticking or defective crankcase breather valve, or a poor seal on the dipstick cap are other common causes of low vacuum and oil leaks.

CRANKCASE COMPONENTS AND FITS

Crankcase Seals

Any device, gasket, or material may be considered a *seal* if it prevents leakage of air or oil between the crankcase and the outside. The joints between the halves of a split-crankcase two-stroke engine, or between the end plate and cylinder block of a four-stroke, are seals. In some cases those may be ordinary gaskets, for example, where a set of external breaker points is mounted on the side of a crankcase. Another example is the small O-ring or rubber bushing around the push rod that operates the breaker points. Still other examples are synthetic rubber strips used between engine sections, or even a sealant compound or paste.

In any case, you can be quite sure that the engine manufacturer has good reason for using the seal method and material he's chosen. Before reassem-

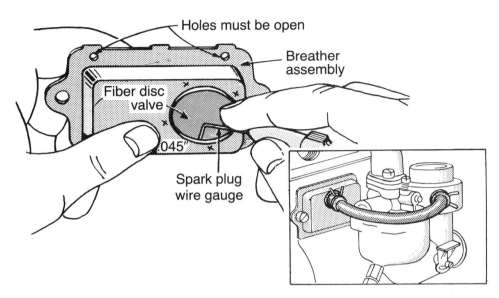

Holes must be open

Breather assembly

Fiber disc valve

.045"

Spark plug wire gauge

Use a wire gauge to check internal clearance of the flutter valve or reed in crankcase breathers.

bling a split crankcase or end plate, be sure to use the proper seal and follow any instructions specific to that engine. If, for example, you have a crankcase split lengthwise through the main bearings, too thick a gasket or too heavy a buildup of sealing material would leave the crankshaft bearings without sufficient clamping action, causing looseness. Too thin a gasket might cause a bearing to be tight in its bore. And if the gasket material is such that it draws down unevenly, it could easily misalign the bearings. Those are the little things that make a big difference in your engine-repair success.

Crankshaft End Play

Most gasket kits for the crankcase end plate will have a selection of gaskets, all of the same shape but of different thicknesses. By varying the combinations of gasket thickness you can adjust the final location of the end plate in or out a few thousandths of an inch from the main crankcase. This establishes a very important running clearance called *crankshaft end play.* The table of fits and clearances found in a workshop manual will specify the correct end play for your engine. Typically that might be .002 to .003 inch for an engine with bushing or ball bearings, though it might range from .003 to .020 inch. Thus,

This crankcase seal was destroyed by improper assembly techniques. Garter spring, which surrounds and pressurizes the seal lip, has been forced out.

with the engine cold the crankshaft can slide back and forth somewhat. As the engine heats up the clearance will probably vary, but it must always be

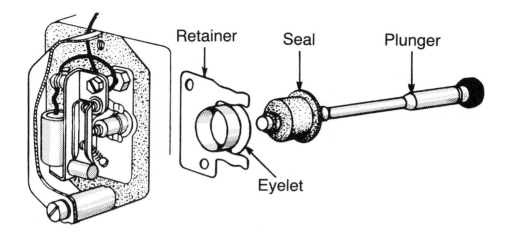

Retainer   Seal   Plunger

Eyelet

This seal keeps oil out of an external breaker box. The seal is a recent addition, although you may buy one for older engines. Be sure to have your small engine's make, model, and serial number handy.

there to prevent the thrust faces of the bearing from having the oil film squeezed out.

When assembling an engine, you can guess at the initial trial-gasket combination or try the one the manufacturer suggests. To make the trial measurement, torque down the end plate retaining bolts as specified to compress the gasket. If the crankshaft is accessible from the bottom, slip a feeler gauge between the crankshaft thrust face and the rear of the bearing after first tapping the end of the shaft lightly to force the shaft fully in one direction. Measure at the end you tapped. A second method requires mounting a dial indicator on the crankcase so the tip bears against the end of the shaft. That provides a direct readout of the end play. Note that some engines do not use gasket thickness to control end play but, instead, use a selection of thin shim-like thrust washers on the shaft.

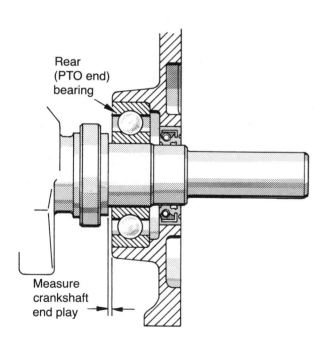

Rear (PTO end) bearing

Measure crankshaft end play

Not all engines have access for measuring crankcase end play directly with a feeler gauge. The alternative is to use a dial indicator on the end of the shaft.

Taper-Bearing End Play

The preceding discussion pertains only to bushing or ball bearings. Tapered roller bearings are often operated with zero end play or even a slight preload or compression. That is because there is no thrust shoulder as such on the crankshaft and the taper of the bearing rollers and races acts to absorb thrust loads. Though each engine manual has specific in-

One way to measure crankshaft end play is to push on the shaft, install a pulley flush against the seal face, and bump the shaft back out. Then measure the clearance with a flat feeler gauge.

structions on measuring the gasket or spacer package for tapered roller bearings, the general procedure is to make a trial installation with the shaft tapped to the end opposite the bearing plate. The cover plate is then put in place and pushed down firmly against the bearing. You'll find a gap between the cover plate and the crankcase, and by measuring this gap with a feeler gage you can determine how thick the gasket must be.

Each of the procedures given here is necessarily basic, since minor variations will be found from engine to engine, even on those of the same make. If they seem a bit involved for installing a simple gasket seal, remember that in addition to preventing oil and dirt leaks, these gaskets are also serving as spacers for major working parts of the engine.

Driveshaft Seals

In addition to the seals at the two ends of the crankshaft, there may also be seals around auxilliary drive shafts such as magneto or power-take-off drives. Nearly all seals, however, are of the same basic construction, although they can differ significantly in design and material.

Examine such a seal and you'll see that it consists of a pressed metal housing or ring-shaped member. Inside the housing, you'll find another ring of soft, molded synthetic material. Literally hundreds of different materials are used for such seals and it's important that the seal you use is the right one. There is no way to tell them apart except by part number. The bore and shaft sizes are pretty much

This crankshaft actually had about .0025 inch end play; when I stopped pushing on the other end, it dropped back almost .001 inch. End play prevents bearing thrust faces from having the oil film squeezed out.

standard except for variations between metric and inch dimensions.

Some seals are intended to resist hydraulic fluid, others engine oil, and still others water or antifreeze. Some are compounded for extreme heat resistance, important in an engine seal. It is also important, in a snowblower engine for instance, that the seal remain effective even in extreme cold.

Internally, the molded material of a shaft seal is formed into a very thin, sharp-edged and delicate lip. In some seals you find two such lips, each curved opposite to each other. In operation, the pressures in the crankcase force the inward facing lips of the seal against the shaft to prevent oil and vapors from exiting. When negative pressure is present, the outer seal lip, curved outward, acts to prevent the entrance

of air and dirt from the outside. Some seals depend entirely on the resiliency of the material to exert a light pressure on the shaft. Other seals have circular springs, called *garter springs,* which augment the pressure of the seal on the shaft.

Shaft-seal removal and installation. The first thing to remember about these seals is that they're very delicate. The thin lips that bear on the shaft are easily broken, nicked, bent, or otherwise distorted. Assume, for example, that you are going to remove the end plate from the engine by removing the retaining bolts and using puller screws to withdraw it over the crankshaft. If you do this without cleaning and polishing as much of the crankshaft extension as you can reach, the seal will certainly be damaged. Even if you take such measures, you can still damage the seal although you may not be able to see the damage. Your best bet is to always replace the seal you removed with a new one.

In many cases you can replace a seal without disassembling the engine. If the seal is at the flywheel end, the flywheel must be removed. Though engine manufacturers do have seal pullers, they are sized to a limited number of engines and you'd be lucky to find exactly the one you needed even if you had a collection. This puller tool is basically a sleeve that fits over the crankshaft extension and has a series of stepped serrations, which slip inside the seal housing and grip the edges. A puller screw is then tightened against the end of the shaft to withdraw the seal.

Lacking such a tool, you can improvise, although considerable caution is necessary.

1. With a thin sharp punch, lightly dimple two marks in the face of the seal housing at 180 degrees to each other.
2. Drill two holes just through the metal at those locations using a small motorized hand drill with a 1/16-inch bit. Avoid touching the shaft or wandering off the face of the seal into the crankcase metal.

Clearance under the bearing retainer is checked, then selected shims or spacers are installed to produce specified bearing pressure.

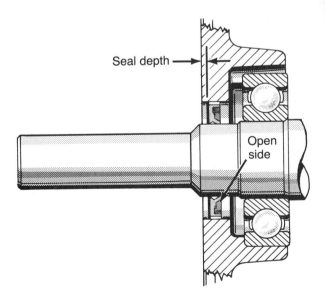

Some seals are pressed in flush with the crankcase, while others are pressed to a controlled depth. Failure to observe this detail will cause interference with other parts later.

3. Thread in two small sheet-metal screws and, working gently from side to side, pry upwards beneath the screw heads to lift out the old seal. *Again—be careful not to nick the crankshaft or counterbore.* Good lighting is a must in this operation.

Installing the new seal requires equal care. Remember the seal lip—or one of them on double-lip seals—projects backwards towards the engine interior. You do not want to have this lip catch and fold the wrong way. Inspect the seal-lip area on the shaft. If it has gum, oil deposits, or grit on it, use something such as a pipe cleaner and lacquer thinner to clean it away. If you can see a groove worn in the shaft where the old seal was running, it is probable that your new seal will not be effective for very long, if at all.

Clean the counterbore in the crankcase as well. In some instances you may find a residue from a sealer with which the old seal was coated. Clean it away and be sure the counterbore walls and bottom are free of nicks or tool marks.

Ideally, you will have protective tools for seal installation. One of these is a sleeve-like device with a slight taper and a very thin edge over which the seal is slipped. A light coat of grease on the tool and shaft will help. By sliding the tool into the cavity until it bottoms you can keep the lip pointed the right way. You must now use another sleeve tool to slip over the shaft and bear against the face of the seal housing. You can then tap the seal home and withdraw the protective sleeve. In most cases, the seal is simply driven flush with the face of the crankcase, at which point it will bottom in the counterbore. Sometimes, however, the seal is driven to a specific depth, and this can only be known by contacting the manufacturer or dealer or consulting the shop manual.

Many mechanics like to coat the outer sides of the seal or the counterbore lightly with gasket sealant. That should not be necessary, but if the counterbore is corroded or scratched, it may help.

As with the special seal-removal tools mentioned

Tapered sleeve tool

A tapered sleeve tool guides the seal lips in this crankcase cover over a step in the crankshaft. Tool is home-made, though a similar one is readily available from a number of engine dealers.

One way to pull a seal without disassembling the engine is to grip it with small sheet-metal screws. The wrench socket, for a fulcrum, and side cutters levered it out easily as shown.

earlier, the protective tool described for the seal lip must be exactly the right size for the seal. If you have a metalworking lathe you can easily make such a tool to size. If not, you might try a technique I've used successfully for many years:

1. Take a scrap of photographic film and wrap it around the shaft to form a tube.
2. Apply some grease to the outside of the film, then slide the seal down the film and into place. The film will slip inside the seal lip and spring out against it, making the job easier.

You'll still need a suitable driving tool to fit the seal face, and it must be long enough to extend beyond the end of the crankshaft so you can press or tap the seal into the counterbore. You can make

one by sawing off a length of scrap pipe or tubing of suitable diameter. Be sure the end that contacts the seal face is square and smooth.

Replacing a seal with the crankshaft out of the engine is a bit easier as a rule. You can gently tap the old seal out from the inside and press the new one in without concern about the seal lip at that time. But when it comes to installing the shaft or the cover plate, all of the preceding precautions must be observed.

TYPES OF BEARINGS

Because bearings are such a vital part of any engine, it is important to understand the characteristics and service procedures for the various types you may encounter.

Obviously, the least expensive bearings for the engine manufacturer are a pair of precision holes bored in the end plate and opposite end of the crankcase. The crankshaft simply runs directly in the crankcase metal. Since aluminum is an excellent bearing material, however, this works out fine for many small engines.

The disadvantage of using the crankcase metal for a bearing is that wear, or any type of damage caused by lubrication failure or a foreign body, involves the entire cylinder block and crankcase. Such damage can usually be repaired by pressing in a replacement bushing, but doing so requires special tools to cut a new, oversize bore to receive the bushing, and later, to ream the bushing to size. The cost of a new engine block may not justify the time and labor involved.

The next grade up in bearings would be a pressed-in bushing of bronze or other good bearing material. Such a bearing, if worn or damaged, may be pressed out and a new one pressed into the original bore. Reaming to final size is still required—again, a special-tool operation. If you're overhauling an engine with either of the preceding bearing types, expect to take the block to a dealer's shop for actual bearing repair unless you are extremely well equipped with precision reamers.

High-quality engines use anti-friction, ball or tapered-roller bearings. Such precision bearings seat in equally precise counterbores in the crankcase and require no reamers or other special tools for replacement. They are critical fits on the crankshaft and are normally installed by pressing or shrinking in place. Removal requires a puller or arbor press setup and replacement may require a pressing sleeve to seat them on the crankshaft.

Needle bearings, essentially miniature roller bearings, are common in small two-stroke engines. Whereas conventional ball and roller bearings have both an inner and outer race, a typical needle bearing may have an outer race or jacket, and the needles will actually roll on the shaft journal. Or, the race may be omitted altogether, and the needles will bear on both the bearing bore and the shaft. Needle bearings are especially well suited to small, lightweight equipment because of their lightness, compactness, excellent high-speed capability, and good adaptation to oil-and-fuel lubrication.

Crankshaft seals can be pressed in or tapped in with a hammer. Above, homemade driver slips inside the seal while flange applies the driving force evenly to seal face. Below, installing tool centers seal while drill press seats it.

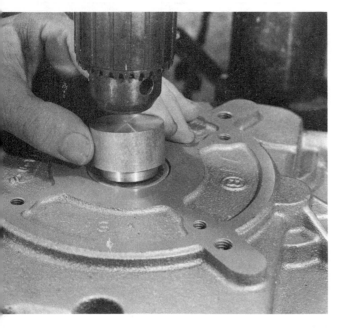

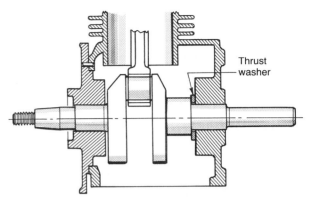

Many light-duty engines use the crankcase material as a bearing as shown above. Replacement bushings (below) are available, but installing, reaming, and aligning them is a shop job.

Photo negative film, well-oiled and wrapped around the shaft, will let seal lips slide home without catching.

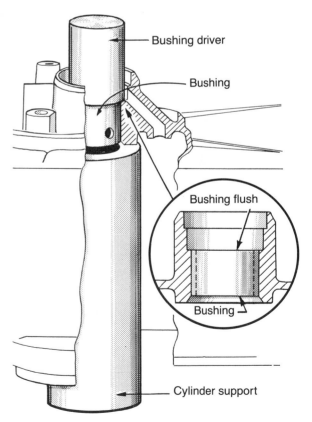

BEARING REPAIR AND REPLACEMENT

I differentiate between repairing a bearing and replacing one. A repair is necessary when the original bearing was simply a hole bored in the cylinder block or end plate. Since this type of bearing is found in the least expensive engines, it always presents a choice of junking the engine, buying a short block, or actually making a repair. If the bearing and engine are badly worn and if other extensive repairs are needed, a complete new engine may be the best choice. If the equipment the engine drives is still in good shape and most of the engine parts are useable, your best bet is probably a short block. And if you are firmly attached to the equipment and want to

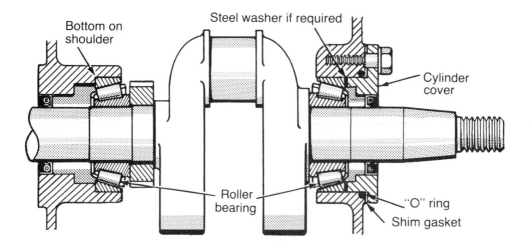

Bottom on shoulder

Steel washer if required

Cylinder cover

Roller bearing

"O" ring

Shim gasket

Tapered bearings above are usually fitted with zero end play. Steel washer beneath the cover plate holds outer race firmly. Ball bearings (below) are normally allowed some end play and tend to adjust themselves.

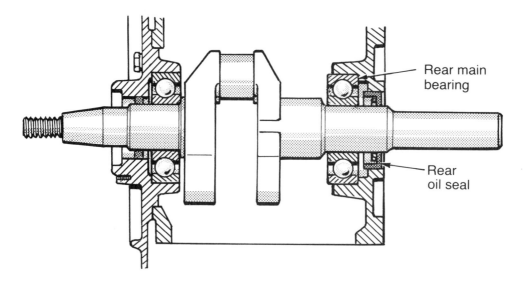

Rear main bearing

Rear oil seal

save the existing engine, repair may be a good choice if your local shop has the tools to do the necessary machine work.

Replacing Bushing-Type Bearings

If your engine has bushing-type bearings, you would not have to bore the crankcase, and the crankcase centers should still be well established. All that's required is pressing or pulling out the worn bushing and installing a new one. A special line reamer is still needed for final sizing, but a good shop should be able to pop in the bushings and ream them in short order if you handle the disassembly and reassembly work.

Servicing Ball and Roller Bearings

Ball and roller bearings are generally used in more expensive engines designed to be repaired rather

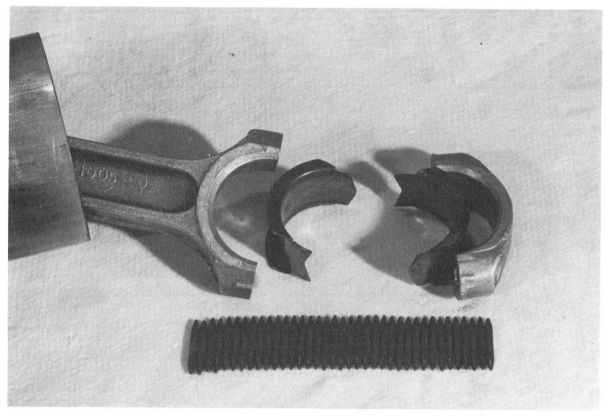

Uncaged needle bearings at bottom in this photo run on two hardened steel shells shown with the rod. Crankpin serves as the inner race. This setup is more difficult to service than the caged-bearing type.

than replaced. Both types are extremely long-lived, and damage usually results from dirt or grit and corrosion. There are some basic points about ball bearings with which you should be familiar.

Although there are many bearings that will "fit" standard shaft and bore sizes, the ones you choose for replacement should either be those listed in the engine manufacturer's parts list, or those purchased from a bearing-supply house that carry an identical bearing number or a matching interchange number. If you examine the bearings in your engine, you'll find a manufacturer's name or code and a bearing number that relates to the type of bearing it is and the job it is intended to do.

Some bearings are intended primarily for radial loads, for example, locating a rotating part such as a crankshaft on center and absorbing a specified amount of side pull such as from a V-belt. Others are built to withstand a certain amount of thrust load in addition to radial load and might be used where a driven device such as a clutch tends to force the crankshaft endways. Still others come with oil shields on one or both sides, felt dust shields, and, typically for electric motors, grease packing. You wouldn't want to use a bearing with a shield that kept lubrication out, and grease lubrication with a sealed and packed bearing is not what you want in a crankcase.

Ball and roller bearing fits. Since there is a difference in the surface areas of the inside and outside diameter of the inner and outer race, the two races are treated differently with respect to tightness of fit. Almost without exception, the inner race will be an extremely tight fit on the crankshaft. The outer race is usually a snug, sometimes called a *line-to-line,* fit in the crankcase and end-plate bores. If the outer

Tapered roller bearings can often be purchased generically by number, though some are specific to a given engine builder. Be sure you have your engine's-make, model, and serial number when ordering.

races were as tight as the inner races it would be very difficult to assemble or remove the end plate or crankshaft from the engine. As it is, puller screws threaded through the end plate are often required to withdraw the plate from the bearing, which remains on the shaft.

Removing and inspecting ball and roller bearings. If you had a shop that was set up to do this job often, you'd probably have a husky hydraulic press and the necessary fixture to hold the shaft and support the inner race while you pressed the crankshaft out. But such presses are rare in home shops and the fixture must be very accurate and solid, so it's hard to cob-

ble together something that will do the job.

The second choice is a puller of the type sometimes called a *bearing splitter*. Such a tool uses two plates that come together to leave a round, central opening. The plates have relatively sharp, tapered edges that can be clamped in under the bearing between the rear face of the bearing and the cheek face of the crankshaft. At the cam-gear end of the shaft you will have to work between the gear and the bearing.

Begin by removing the crankshaft from the crank-case. Although the procedure varies slightly from engine to engine, the general procedure is shown in

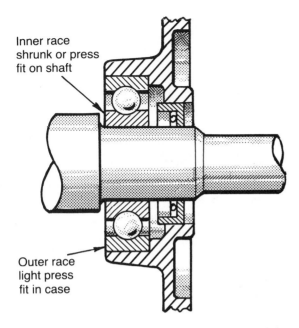

Inner race shrunk or press fit on shaft

Outer race light press fit in case

Ball bearings are usually fitted tightly on the shaft by pressing or shrink-fitting.

the accompanying photos. Once the shaft is out, you'll be exerting enormous pressure against the shaft end by tightening a puller screw against it. To avoid damaging the crankshaft end—especially if the end is tapered down and threaded—drill a shallow centering dimple in the exact center of the tip to center the puller screw. Use this as a seat for a centering button, then lubricate both the button and screw tip and proceed as follows:

1. After initial take-up on the puller screw, stop and examine the puller installation to be sure it is pulling squarely and evenly on both sides and that the puller screw is aligned exactly with the shaft.

2. If the bearing race doesn't come off with moderate pulling tension, take a second look at the shaft and bearing to be certain there are no small wire- or snap-ring retainers that haven't been removed. The wire kind may be hard to see.

3. If not, bring the puller to maximum pressure and give the top, or face, of the puller screw a smart whack with a hammer. *Do not do this with the opposite end of the crankshaft clamped in vise jaws or you risk*

springing the shaft out of alignment. Confine clamping and support to the roughly finished part of the crankcheek at the end where you're working.

4. If puller tension softens perceptibly with the hammer blow, the bearing is moving and the process needs only repeating until the bearing comes free. If you feel no softening of puller tension, apply heat with a propane torch or the like.

5. Confine the heat to the bearing and race since you want to expand it, not the crankshaft. Apply heat while maintaining puller tension until the bearing is hot enough to melt solder wire touched against it. That is about 430°F and should be enough to loosen the bearing with more pressure and tapping.

Don't be surprised if the bearing loosens with a sudden lusty pop before reaching the elevated temperature. Sometimes it only takes a little heat. Once the bearing is loosened, you can usually keep it moving until it comes off. Never reuse a bearing after heating and pulling. If the bearing can be slipped off easily, you then have to decide whether the bearing is really in poor shape. The only practical way to do this is to wash the bearing completely free of oil and dirt as follows:

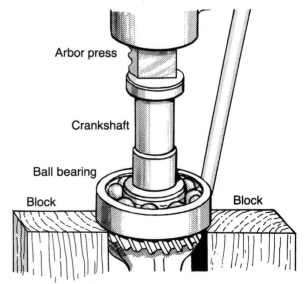

Arbor press

Crankshaft

Ball bearing

Block   Block

One way to remove a crankshaft bearing is with an arbor press and suitable support blocks, though a husky puller with a backup plate is safer.

Removing A Crankshaft

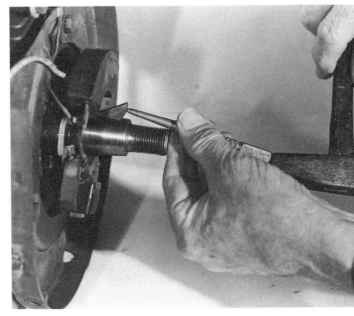

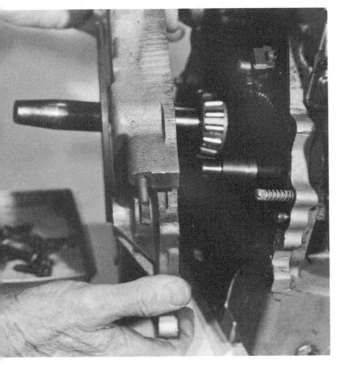

1. Begin by removing the flywheel. Use a puller for a heavy wheel like this one.

2. Don't overlook the flywheel key—tap it out carefully to avoid damaging key or keyway.

3. With cover-plate screws removed, slowly remove the cover. Be sure camshaft doesn't come with it.

4. Draw camshaft straight out. Then remove connecting-rod nuts, rod, piston, and crank.

Don't even try to pull a bearing such as this without a sturdy puller. Expect to damage the bearing.

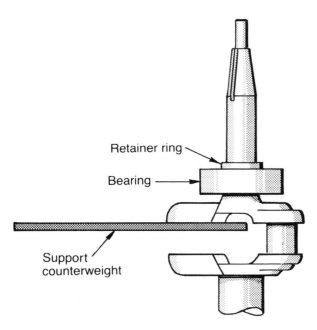

Retainer ring

Bearing

Support
counterweight

Here, bearing is removed using a press and support plate. Avoid putting a bending load on the shaft.

1. Apply solvent vigorously with a stiff, clean brush until the solvent runs clear.
2. Dry the bearing with compressed air if available, but do not spin the bearing while doing so.
3. Rotate the outer race with your fingers while holding the inner race. Any obvious roughness or hint of roughness is cause for rejection.

If you spin the race with your fingers and it sounds noisy or rattles, it needs replacement. This may be normal wear but there's a good chance the bearing is Brinelled. Also examine the outer-race seating surface carefully. You may be able to spot corrosion or rust, or a gum or varnish deposit. Take it as a lesson, and before installing a new bearing, clean the counterbore thoroughly and give some thought to better storage practices and more frequent oil changes in the future.

Replacing a ball or roller bearing. Once a bearing has been pulled it should be scrapped. The pressure required to install a new bearing with an arbor press may be quite high and, again, the support require-

With the rollers stripped away, torch heat is applied while inner race is under maximum puller tension.

ment exceeds anything likely to be found in your home shop. The preferred method is to expand the bearing with heat. Even so, you must be prepared to go about the job with great dispatch.

1. Begin by polishing all gum or rust from the shaft bearing area, also making certain there are no burrs or nicks.
2. Clamp the shaft by the crankcheek on the end you're working on and be certain the grip of the vise will withstand some pressure. The shaft end should be pointed upward. It also helps to smear some lubricant on the area where the bearing race must slide.
3. You'll need a driving tool such as a piece of pipe or heavy tubing that will just slip over the crankshaft and bear against the inner bearing race. Often the bearing will simply drop into place with a gratifying click, but a quick final tap is always needed to be sure

it's fully seated. Have the driving tool and a hammer handy because you'll only have a second or two before the bearing contracts and locks on the shaft.
4. Meanwhile the bearing can be heating in hot oil. A tin can partially filled with engine oil and heated on a hot plate will do nicely. Bend up a small trivet support from screen wire or the like to hold the bearing up off the bottom of the can. Also make a hook from welding wire or a coat hanger to fish the bearing out of the hot oil.
5. Heat the oil until it is smoking and leave the bearing in until it is thoroughly heated; ten minutes is not too long. Then fish out the bearing and drop it flat on some paper toweling or wrapping paper.
6. Take one last look to be sure the proper side is out, if applicable. Then—using welder's or kitchen hot-pot gloves that resist heat

Brinelling on outer race

Common wear pattern in anti-friction bearings is called *Brinelling* after the metal-hardness test that forces a hardened steel ball into a test surface. Such a bearing must be replaced.

and keep the hot oil from soaking through—grasp each side of the bearing and drop it as squarely as possible over the shaft. With good luck, the bearing will clunk into place.

7. Follow instantly with a tap from the driving tool and hammer and let the bearing cool.

If the initial placement wasn't quite square, try squaring the bearing with the driving tool and tapping it home. If it sticks partway down the shaft, do not continue trying to drive it down. You have no choice but to pull it and try again. In extreme cases, try putting the crankshaft in your home freezer to contract it slightly.

Replacing Needle Bearings

If you have never worked with needle bearings before, your introduction to them may be quite frustrating. The caged type with an outer race or shell is seldom a problem since the cage prevents them from falling loosely. Usually the cage and bearing

New bearing was heated for about 20 minutes in smoking-hot engine oil and quickly dropped over the shaft.

Homemade driving tool shown contacts only the inner-race face.

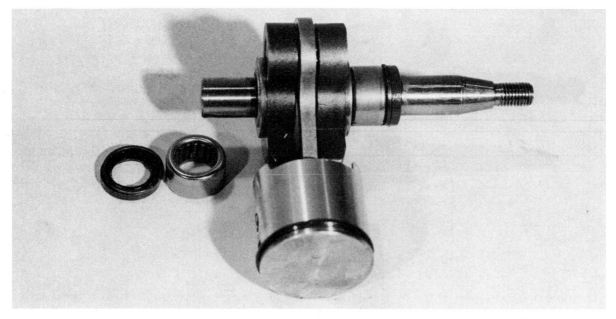

This caged needle bearing and its seal are easy to handle because there are no loose needles.

can be pressed or driven in or out with a tool made to bear squarely against the cage end. Most instruction manuals advise you to drive or press on the end with the lettering, at least when installing new ones.

Although installing such bearings is fairly easy, you should first examine the surface of the crankshaft or other shaft on which they run. If a good light and a magnifying glass reveal a series of longitudinal light grooves or pits, the needles have worn or impressed themselves into the shaft. Such a Brinelled shaft must be replaced since new bearings will not run on it satisfactorily.

Handling uncaged needle bearings. Uncaged bearings can be more exciting. They are simply loose needles or rollers, which are easily dropped, lost down in the engine, flipped out onto the shop floor, or otherwise prone to misbehavior. Moreover, at reassembly they must be packed and held around a shaft while another part such as a connecting rod is slipped over them.

The best way to avoid problems is to anticipate that you'll drop a few, or that they will all fall out randomly. Before starting, place the engine in a shallow pan or box and then pack the area under the bearing with a clean catch cloth. Each bearing takes a certain number of needles, for example, 23. Since you probably won't know what the number is until you have them all out and counted, one lost along the line can spell trouble at reassembly. One favorite hiding place for a stray needle is the bypass passage in a two-stroke.

When you buy a new set of needles for an uncaged bearing, the proper number will be lightly adhered to a paper tape. They will also have a coating of wax that will stick them to the crankpin or wherever else they have to go. Before placing them around the shaft, clean the shaft with lacquer thinner or the like so it is dry, not oily. Oil will dissolve the wax and they'll fall off while you're working. The trick is to place them neatly around the shaft, being careful that they're fully parallel to the shaft and each other, and then slip the assembly together.

If you're replacing used needles, the suggested technique is to clean the shaft and the needles, count the needles to be sure they're all there, and coat the shaft with sticky grease to which the needles will adhere. That works most of the time, but if your grease doesn't seem sticky enough, you might try melting a small amount of beeswax and rolling the needles in it until it cools and becomes adhesive. The wax will stick to a clean, dry shaft while it is slightly warm. After assembly, lubricate the new or used bearings freely with clean engine oil to dissolve and wash away the wax or grease.

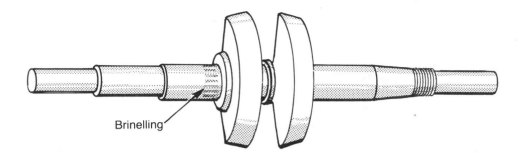

Brinelling

Needle bearings that run directly on the crankshaft can also Brinell and produce a rough surface.

To reassemble a connecting rod with loose needle bearings, use sticky grease in the rod and cap to hold each one as you place it. Tweezers help, as does plenty of care and patience.

16

Crankshafts, Camshafts, and Oil Pumps

In small two-stroke engines the few parts inside the crankcase are the crankshaft, connecting rod, and reed valves if the latter are used. But four-stroke engines must also have at least a camshaft and its drive gears, perhaps a governor, and a set of balance weights, and often an oil pump of some sort.

CRANKSHAFT INSPECTION

The first step in any major overhaul, after disassembly, is a step-by-step inspection of the crankshaft both for condition and dimension. Bent crankshafts are fairly common on direct-drive rotary mowers. They are less common in engines that drive through V-belts or other means. If your rotary mower has struck an obstacle, the bending may be so bad that you can actually see it, or it may show up as heavy vibration.

Less obvious bends of a few thousandths of an inch cannot be detected without precision shop facilities that allow you to mount the shaft in V-blocks or between pinpoint centers for dial indicating. Unless you have the necessary test equipment and are aware of an accident that might have bent the shaft, such slight irregularities probably shouldn't concern you.

Initial Inspection

Begin with a visual check for gross defects. There is little point in making fine micrometer checks only to find out later that the keyway slot is hopelessly battered.

Look for evident signs of distress. Are there discolorations on the crankpin or crank journals, which suggest overheating or lack of lubrication? Can you see or feel grooves in the bearing or shaft-seal areas? Do the keyways appear battered or worn? If there are flats to operate a breaker system or fuel pump,

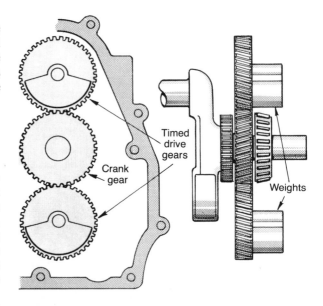

Weights, timed to develop forces opposing those inherent in the crankshaft, piston, and connecting rod mass, are driven by various gear arrangements depending on engine make.

are they worn or grooved? What about the threads on threaded shafts—are they suitable for cleanup to safe condition? Are the camshaft-gear teeth chipped or worn?

Inspecting Crankshaft Dimensions

If the shaft passes initial inspection, it's time to use your micrometers to measure the crankpin and journal dimensions. Measure at several points to detect taper or lack of parallelism along the pin or journal. Repeat these measurements at 90 degrees to the first series. You can then determine if the crankshaft is acceptable or not based on the specifications for your particular engine. In engines with needle

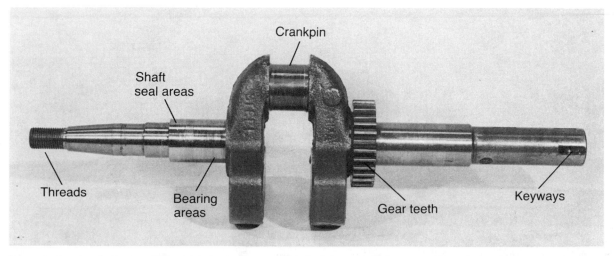

Be sure to check these vital crankshaft areas. This shaft is scored from running without oil.

bearings, either wear beyond limits or signs of Brinelling are cause for rejection.

Undersizing. In many four-stroke engines, especially those intended for heavy duty service, it may be possible to have the crankpin reground to a .010- or .020-inch undersize. That means you'll need a new connecting rod or connecting-rod bearing of equivalent undersize. Regrinding is a highly specialized machine-shop procedure and requires specific fixtures to mount and spin the shaft about the crankpin center. Your local engine service shop or parts agency may not be equipped for the job, though they may have a source that regrinds small shafts. Thus, you may be able to buy a reground shaft and connecting rod at a substantial savings compared to a new shaft.

CAMSHAFTS

Although a small engine camshaft is a simple device in itself, it can be one of the more awkward parts to remove and replace because of its location relative to the crankshaft, the need to match specific teeth on assembly, and the control of the running clearances and end play in the internal bearings of the crankcase.

In addition, many engines are equipped with automatic compression releases to reduce cranking loads, and some may even have power-take-off drives or clutches off the camshaft to provide a drive at half crankshaft speed.

In the simplest arrangement of the crankshaft and camshaft, removal of the end plate will reveal the crankshaft gear and camshaft gear neatly meshed. Turn the crankshaft until the marked teeth on each gear index with each other and you can immediately see if the cam is properly timed. To remove the camshaft you merely lift it up and out.

Removal and replacement become more difficult if the engine has a ball bearing on the crankshaft above the cam gear. Now, when you attempt to lift up the cam gear it will strike the ball bearing and cannot be removed. This means that the crankshaft must come out and that entails removing the flywheel at the opposite end. It also means that the crank and the camshaft must be inserted in the crankcase together with the gears meshed in time. The timing marks may be hard to see because of the bearing, or, they may be located elsewhere, perhaps on the crankshaft counterweight. Be sure you've located the marks before disassembly.

A further complication is involved in engines with the crankshaft and camshaft gears at the end of the crankshaft opposite the removable end plate. This puts the cam gear back inside the crankcase and intrudes the counterweights in the line of vision and the path of removal. Such camshafts are usually supported by a shaft that extends through the crankcase wall. To remove the camshaft, the support shaft

A micrometer check will tell you whether a crankpin is out-of-round or tapered.

must be driven out to allow the cam gear to drop down into a casting pocket for clearance. None of the arrangements is especially complicated if you follow the general removal steps included here and supplement them with a few specifics from manual for your particular engine make and model.

Camshaft Inspection

Inspect a camshaft much as you would a crankshaft. Examine the bearing ends of the shaft for signs of burning, bearing or lubrication failure, and Brinelling in the case of needle bearings. Assuming those are in good condition, the critical parts of the shaft—in addition to the drive gear—are the cam lobes.

If the lobes appear smooth and polished there is probably no problem. Checking them with micrometers against the manufacturer's specifications will show if they're worn. Wear or signs of scuffing, scoring, or chipping are causes for a new camshaft. Generally, it isn't economical to try to reshape small-engine cams such as might be done on a larger engine. If the engine is an antique or collector's item, you might pay to spray or weld on molten metal and regrind a cam, but that is a specialty-shop job. In all cases, if the camshaft gear is damaged, both it and the mating crankshaft gear must be replaced.

Valve tappets. Whenever you withdraw or install the camshaft, the tappets, or valve lifters, should be pushed up out of the way so they don't catch on the cam lobes. When removing the camshaft, try to have the engine inverted, or at least on its side so the

Inspecting Valve Lifters

1. Intake and exhaust lifters are often identical, though not always. It's best to put them back where they ran originally. This is especially true if an automatic compression release is used.

2. Lifter faces should be mirror-smooth and bright. Circular scratches here denote common practice of locating lifters slightly off cam-lobe centers to impart a rotating motion for even wear.

tappets don't drop out by gravity. Then, remove each tappet only after ascertaining whether it is for the intake or the exhaust and marking them. There are two reasons for that.

First, although they may appear identical, they may not be twins. This is especially true if an automatic compression release is used. Here, the exhaust tappet may be shorter than the intake. The second reason is that it is good mechanical practice to make sure parts that have run together, go back together. Remember, they have inevitably worn slightly to fit the cam lobes and there is no reason to upset that.

The faces of the tappets should be highly polished and very smooth in the cam-lobe contact area. Examine them with a good light and a magnifying glass to spot any pitting, scratching, or chipping. In general, the tappets seldom need replacing unless the cam is also in trouble.

Camshaft Timing

Nearly every engine has definite marks on the cam gear and crank gear, or on a related part such as a counterweight, which must be matched on assembly to locate the cam lobes so the valves will open and close at the proper time. Sometimes these marks are readily visible and at other times, particularly in very old engines, you may have to search for them. Occasionally, instead of a stamped mark or a dim-

Typical camshaft drive gear meshes in timed relationship with crankshaft gear. Mushroom-shaped valve lifters ride against cam lobes. To remove the camshaft, you can often simply lift it up and out.

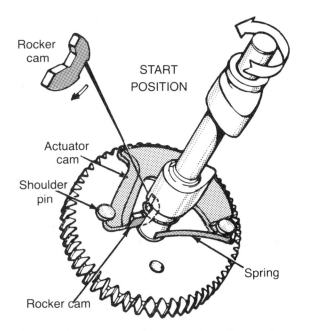

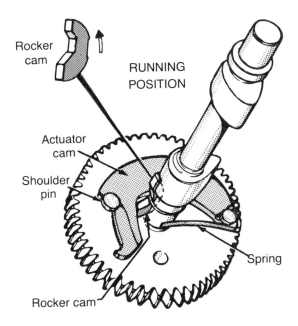

In typical cam-mounted compression release, above, a moveable member holds the exhaust valve open past TDC to ease starting. As the engine starts and runs, centrifugal force deactivates the device.

On many engines, valve timing requires simply matching the marked teeth of the cam and crankshaft gears.

ple, you'll find one gear tooth with a corner beveled. One major manufacturer simply uses the keyway for the crankshaft gear as a timing mark.

In such cases where the mark is doubtful, you should make some checks before disassembly and make your own marks if necessary. After assembly, even if you're confident that you've matched the marks, it always pays to check; it's easy to slip a tooth or two, particularly with helical-cut gears. To make the check:

1. Slowly turn the crankshaft in the direction of normal rotation. If the valves and springs are out of the engine, you'll simply be moving the piston.

2. Use a flashlight if necessary to observe the cam lobes. As you see the intake lobe start to contact the tappet, the piston should be moving towards and very near top center.

3. Place your finger on the intake tappet and press down so you can feel the initial lifting action of the cam. If the cam is timed properly, you'll normally feel the lifting action at top center or just before. Remember—with the valves in place, the tappet will

have to move up the distance provided for valve clearance before the valve will actually start to open.

Since the crankshaft and camshaft gears have normal backlash, or some extra backlash from wear, it is important to maintain a constant pressure in the drive direction to make up for the slop in the gears.

If the intake tappet is actuated near top center you can feel that the timing is probably right. If you are disassembling an engine on which you weren't sure of the marks, now is the time to mark three gear teeth that are meshed—one mark on the center tooth and one on each side.

For a further check, continue to rotate the crankshaft and feel for intake closing when the tappet bottoms on the base circle of the cam lobe. This should occur near bottom center. The next 360 degrees of crankshaft rotation should show no tappet lift since these are the compression and power strokes. At the bottom of the power stroke you should start to feel the exhaust tappet lift, and it should remain lifted until it gradually closes at about top center.

If your engine has an automatic compression release, expect to see some exhaust-valve action during the compression stroke. This can be ignored. If everything happens pretty much as scheduled, you can be reasonably sure you haven't gone astray. One note of interest if your engine was built for certain models of Sears equipment: you'll find the cam-gear timing set one tooth ahead of the usual position, to your right. Reassemble it to the same setting.

AUTOMATIC COMPRESSION RELEASES

In Chapter 14 I mentioned that some two-stroke engines have automatic compression releases to ease starting. These use a reed-type valve since the combustion chamber has no poppet valves. In many four-stroke engines the cam gear is fitted with one of several counterweight-and-spring devices to relieve compression somewhat during cranking and allow manual starting of some larger engines.

Although the physical parts of these devices

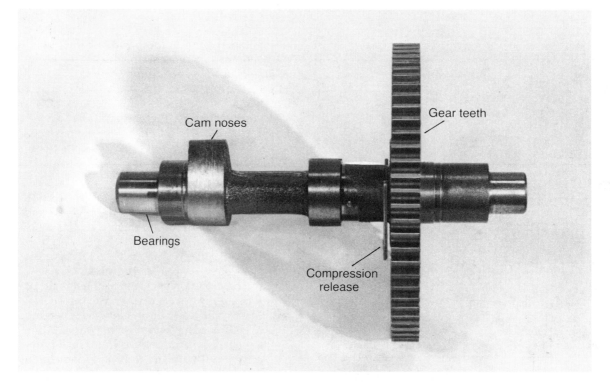

Cam noses

Gear teeth

Bearings

Compression release

Vital camshaft dimensions to check include those of end bearings and lobes as well as gear teeth.

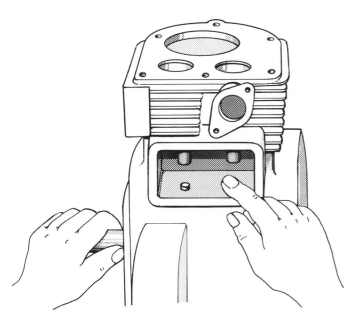

The intake valve should just start to open at or closely before top center, and the exhaust lifter should close at about bottom center except on an engine with an automatic compression release.

differ, they generally work on the same principles. At rest, the weight or weights are held by the tension of a spring so they act to actuate a tab or pin on the exhaust cam. As the engine starts and increases speed, the counterweight is moved out by centrifugal force and the compression release is inactivated. In the release mode the exhaust valve is held open for a short while during the compression stroke to reduce the amount of air trapped above the piston.

Most of these devices are designed to last the life of the engine and, unless damaged by rust or corrosion, seldom need replacement. In most cases the automatic compression release cannot be repaired, although some have provisions for replacing the springs. The main checks you should make are for free, smooth, and easy movement of the parts. If the weights operate freely, the tab or lobe moves nicely, and the springs return the weights without hesitation, there is probably no reason for concern.

At least one maker, Kohler, does recommend checking the timing of the compression release relative to piston position. This is done with the cylinder head off and the valve lifted by the compression release. If the valve height is incorrect, adjustment is made by slight bending of the release tab.

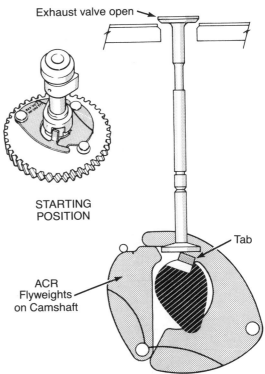

This version of an automatic compression release holds the exhaust valve open during the beginning of the compression stroke.

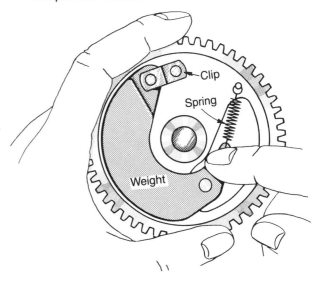

Weight and springs in an automatic ignition advance are similar to those in a compression release. But this device actuates breaker-system advance and retard.

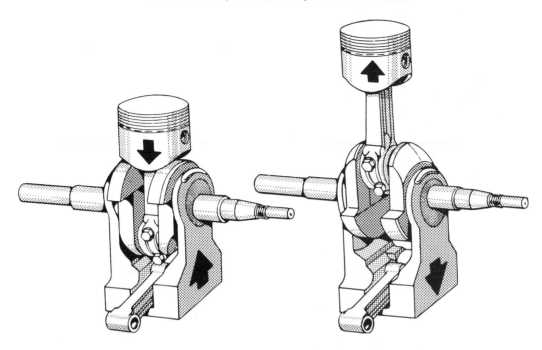

Oscillating balance weights serve the same purpose as the rotating type, but shuttle up and down opposite to piston travel. These devices tend to cancel out the inherent roughness of single-cylinder combustion.

AUTOMATIC SPARK ADVANCE

In at least one make and model series you will find a similar weight and spring device that is unrelated to the valves, yet adjusts the ignition timing relative to engine speed. The principle is much like the automatic advance weights in an automotive distributor. Again, free movement and a spring with enough tension to lift the weight is all you need to look for. These parts can be freed up or replaced if necessary.

CRANKSHAFT BALANCING SYSTEMS

A single-cylinder engine tends to be inherently rough running, and a fairly massive flywheel is needed to smooth it out. Although most engines have crankshaft counterweights to counterbalance the piston and connecting rod to some extent, many engine users still find the roughness or vibration unpleasant. Thus, many engine manufacturers provide moving masses, either rotating or oscillating,

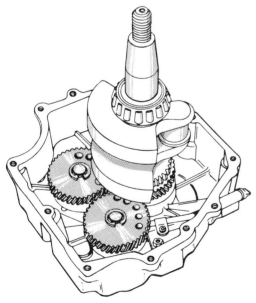

Rotating balance weights may be located at either end of the crankshaft. In addition to timing, be sure the bearings—usually needle-type—are running free and in good condition. Also inspect gear teeth.

which are timed to exert their forces in opposition to the natural unbalanced forces of the engine.

The trade names of those systems may be Dynamic Balance, Dyna-static Balance, Synchro-Balance, and so on. At least three of the more common systems use rotating weights driven by the crankshaft gear but rotating in directions opposite the crankshaft rotation. The weights and gears may be located on the crankshaft itself, at opposite ends, or on support shafts above or below the crankshaft.

Oscillating balance weights have eccentric journals inboard of the main journals at each end of the shaft. The weights are mounted on these eccentric journals and move up and down as the eccentric turns within them. A third member called a *link,* which resembles a small connecting rod, is pivoted on a support pin in the crankcase and on a dowel pin extending between the two weights. The link stabilizes the weights so they travel only in a semivertical direction and do not rotate.

Timing Balance Weights

The whole principle of ironing out the off-balance impulses with moving balancers is one of timing. Each manufacturer has his own method of timing the balance weights. The only system not requiring timing is the oscillating type since the eccentrics on the crankshaft are a fixed part of the shaft and no gear teeth are involved.

Two other systems start the balancing procedure with the piston at top center. It can be either the end of the exhaust stroke or the end of the compression stroke. Briggs & Stratton uses short lengths of $\frac{1}{8}$-inch rod, such as brazing rod, inserted through normally plugged holes in the engine end plate. The end plate is assembled and the gears meshed with the locating pins in place. The small sealing plugs are then replaced. Tecumseh also provides plugged observation openings, but has slots in the balance weights for visual reference through the openings. A screwdriver or the like is used in the slots to adjust the gears for meshing and alignment during installation of the end plate.

The timing procedure used by Kohler for the Dynamic Balance system is slightly more complicated and is best performed using a timing tool that meshes with the gear and holds it in position while meshing the crankshaft drive gear. An alternate procedure without a tool involves visual matching of a

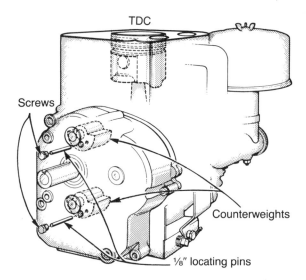

On this engine, two small screws are temporarily removed and pieces of one-eighth-inch brazing rod inserted in holes in the weights to secure them against accidental displacement.

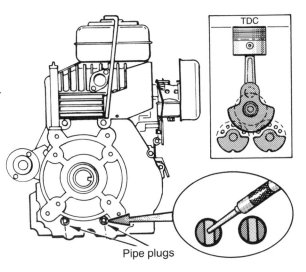

This engine has two pipe plugs that let you see the timing slots in the weights and, if necessary, poke the weights into position so they line up with the holes.

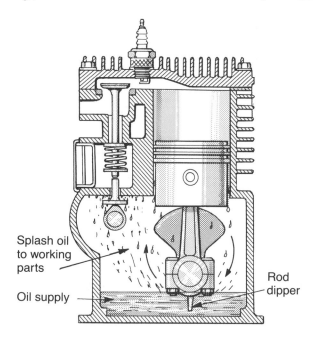

Splash oiling is the norm on horizontal-shaft engines. Vertical-shaft engines require some other form of oiling since the oil level is below the connecting rod and upper bearing.

series of marks on the crankshaft and two balance gears. Since several end-play and running-clearance checks are made at the same time, a shop manual is definitely needed.

LUBRICATION SYSTEMS

The most common oiling method for small gas engines is simple *splash lubrication* using a small, finger-like dipper extending down from the bottom of the connecting rod. Often a trough-like reservoir is provided and the dipper passes through the trough on each revolution of the crankshaft. If that sounds like a rather haphazard method, remember that many hours of engineering and testing go into working out the proper size and shape of the dipper. Moreover, the dipper is placed and angled to accurately direct the oil to critical surfaces. Therefore, be sure to observe and note the position of the finger on disassembly. It's often possible to assemble one backwards. Also, if the dipper is a pressed metal stamping, examine it carefully for signs of cracking.

If a dipper breaks off while you're using the engine, perhaps from vibration, the engine will soon lock up and will often be severely damaged. By the same token, this type of oiling system is vulnerable to acute operating angles that do not allow the dip trough to fill properly. Bear this in mind particularly when operating a mower up or down steep slopes.

A second common means of oiling is a centrifugal slinger, or small rotating paddle wheel, which drives off one of the internal gears and throws off the right amount of oil in the proper direction. Things to check here are wear on the driving teeth, broken paddles, wear in the slinger-hub bearing, and the location and placement of any oil-deflector baffles. Sometimes certain areas may be exposed to an excess of oil, and small metal barriers or deflectors are used to redirect it.

Pumped Lubrication

Many popular small engines use some form of positive-displacement oil pump. Such a pump may consist of a simple plunger and barrel driven by an eccentric or flat on the camshaft, or it may have

In this vertical-shaft engine, a plastic slinger drives off the camshaft and throws oil to selected points.

This small oil passage carries oil to the upper crankshaft main bearing. A bit of carbon, dirt, or even lint from a wiping rag could plug this hole and ruin a perfectly healthy engine.

lobed rotors working in eccentric lobed chambers and driven off the crankshaft gear.

In both cases, the oil is conducted through drilled passages, usually starting with a drilled camshaft, and distributed to the bearings and other areas by such passages or spray outlets. This is a common arrangement on vertical-shaft engines, since the lower bearings are bathed in oil but the upper crankshaft and camshaft bearings are out of the oil at the top of the engine.

In some cases oil, under pressure, will be transferred through a hole in the upper main bearing to a drilled passage leading to the connecting-rod bearing. This type of oiling system requires a little more attention during repair. First, you must be certain with the plunger-type that the drive eccentric is not worn and that the plunger moves freely and smoothly. All parts, including plunger, plunger barrel, and drilled passages, should be cleaned and freed of gum or deposits. The small intake screen should be removed and any sludge trapped beneath washed away with solvent.

Go through the drilled passages of the entire system with a brass rod or small bristle brush. Remember, although this is a pressure system much as you have in your car engine, there is no oil filter except on larger industrial-level engines. Thus, even small particles of carbon, lint from wiping rags, grit from careless oil handling, or oil sludge are not filtered out but can find their way to a critical point and either block oil flow or damage polished running surfaces.

As a final step before closing up an engine with a positive-displacement oil pump, squirt plenty of oil into the pump pickup area to be sure the pump is primed. Follow that by hand-turning the crankshaft to initiate the pumping process. You should be able to see oil discharge from some of the drilled passages or the connecting-rod bearing. This will eliminate the possibility of an air lock where the pump operates in a bubble of air for some time before it picks up oil. Because your small engine does not have an oil-pressure gauge or warning light, you could run without lubrication long enough to do serious damage without realizing it.

The ball end of the plunger-type oil pump above is seated in the cover plate (below), and the eccentric on the cam causes the plunger to reciprocate in the bore. Passages in the cam act as inlet and discharge ports.

17

Pistons, Rings, and Connecting Rods

When you consider that a four-stroke engine running at 3,600 rpm subjects the piston to 1,800 fiery explosions per minute—30 per second—and that the piston is repeatedly being pulled and tugged to reverse its direction twice in each revolution, it is surprising that pistons endure as well as they do. About the only saving factor is the lubricating oil that splashes against the interior of the piston crown and skirt to carry heat away.

PISTON VARIATIONS

If you carefully read the shop manuals for small engines, you'll find that specific piston materials and part numbers are listed for specific models. Some pistons will have tin plating, others will have chrome plating, and still others will have no plating. They will also vary in the number of ring grooves.

Many engines also have the connecting rod offset so its center does not align with the center of the piston. Thus, although the piston may appear to be the right size, it probably won't run satisfactorily in your engine unless it is the exact one specified in the parts list for both the engine model and specification number and sometimes, even, the serial number. Often a manufacturer will make design changes in a basic model or incorporate modifications for a given customer.

Piston markings. Although it is tempting to just clean up the ring step at the top of the cylinder bore and push a piston out of an engine, it is wise to first do some searching for marks that tell you how the piston goes back and perhaps what type and size it is. Clean the combustion deposits from the crown and use a good light. Pistons that are .010-inch oversize, for example, will usually be so stamped. Look next for markings indicating which way the piston goes in the bore. Such marks may be a small arrow, notch, or dimple on the front, right, or left side of

the crown. If you plan to reuse the piston, it doesn't hurt to make a light punch mark yourself.

You may find other marks that can be evaluated only by checking a manual or asking the dealer or manufacturer. For example, Tecumseh stamps the letter L on tin-plated pistons to be used in sleeve-bore engines. An apparently identical piston, but without the letter stamping, is chrome-plated and is to be used only in aluminum-bore engines.

Other piston markings can be seen only after the piston is out or almost out of the cylinder. Some of the markings, such as casting numbers, are incidental to engine manufacturing, but they're often a convenient locating reference for assembly. Thus, the manual may advise you to assemble the piston with

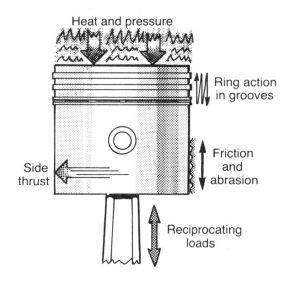

Every time your engine fires, the piston is impacted by tremendous heat, pressure, and friction. Detonation is often more than the piston can stand.

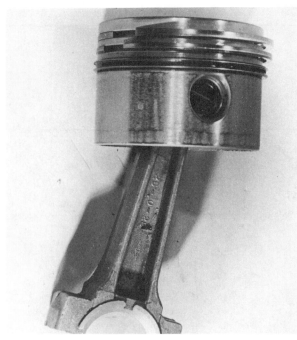

Arrow above shows the correct direction for installing the piston. If you don't find such a mark on the piston crown, sketch or note features such as casting marks on the connecting rod (left).

the casting number on the inside of the piston skirt oriented to the camshaft side. Without a manual, you'll have to rely on careful observation and your own markings.

PISTON REMOVAL AND INSPECTION

Two types of connecting rods are commonly used on small engines. One, routinely used in four-stroke engines, has the lower or *big* end of the rod split to form a cap that is mated to the upper section of the rod with capscrews or nuts. The other type is not split but uses needle bearings. To remove the connecting rod and piston in the latter type of construction, the crankcase end plate must be removed and the crankshaft, together with the needle bearings, slipped endways out of the rod to allow the rod big end to be maneuvered over the shaft as the shaft is withdrawn.

On other engines using a needle-bearing connecting rod with solid big end, the cylinder itself is unbolted from the crankcase and withdrawn, leaving the piston and rod still in place. Those engines are built with a blind-end bore, and the cylinder and head are integral—a design typical of small two-stroke engines such as those used on chainsaws. Again, it is very important to look for markings on both the rod and piston. If you can find no meaningful marks such as a casting number, make some light ones of your own and make a little sketch to help you on reassembly.

Rod-cap matching. Almost without exception, you'll find some type of markings to show how the rod upper half and the cap should be matched. These may be notches, dimples, stampings, or any of many marks, but they will tell you that the rod must be matched to the cap in a certain relationship. The marks may also indicate that a given face of the rod must face out or in on assembly.

Some manufacturers separate the rod and cap by fracturing it in half. That leaves the mating faces with small irregularities, which can be matched only one way and which realign the rod cap perfectly. Be sure you note the arrangement of the oil dipper, small lock plates, tabs, or other pieces on the bottom of the rod cap. Again, it pays to make a sketch of the exact location and positioning of such parts. If you err here, you may reassemble your engine so it won't turn over, won't lubricate, or makes unfavorable noises when running.

Connecting-rod bolts. The very strong forces exerted on the connecting rod and cap mean the bolts retaining the cap must be extra-high-grade, properly torqued, and securely locked. Most modern engines will have some type of locking nut or capscrew. One of those is the type of nut using a plastic or fiber insert or collar to exert constant pressure on the threads. Treat such nuts as expendable and use new ones on reassembly. After many hours of running in hot oil they may well be less effective in their locking action.

Another type of fastener will not accept a standard hex wrench. If you encounter such nuts, you'll have to go to an automotive store and buy a socket made especially for them. Still another type has small serrations on the contact face and locks by slightly distorting the head.

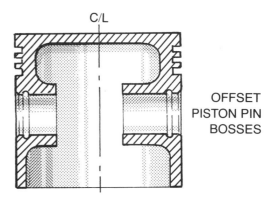

OFFSET PISTON PIN BOSSES

Just because a piston will fit in a cylinder bore doesn't mean it's the right one. The pin bosses may be offset as shown here, or the material or surface finish may not be suitable for your cylinder-bore material.

Also record the location of whatever marks you can find on the piston and relate them to the rod markings.

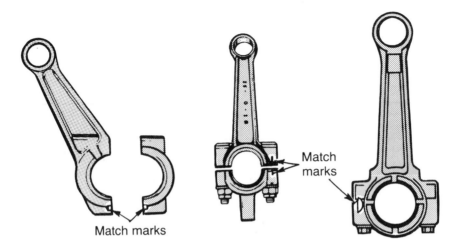

Match marks

Match marks

Match marks

Split-end rods come in many configurations and may be split horizontally or diagonally. All must be assembled with rod and cap in the correct relationship, or your engine won't run for long.

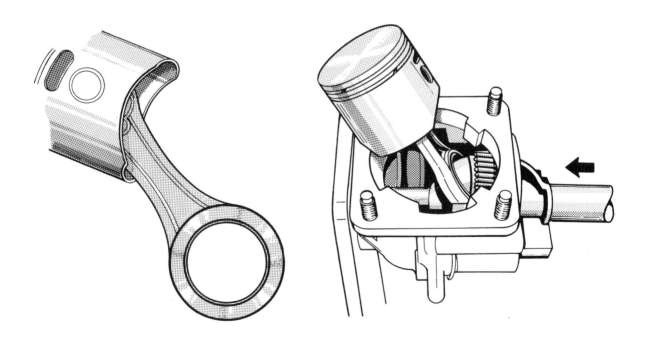

A common design in small two-strokes is this unsplit lower rod bore, used as the running surface for the needle bearings. Removal requires maneuvering the rod bore off the needle bearings as shown.

Two-stroke cylinder, shown being removed, is typical of engines where the crankcase is part of the machine.

Probably the most common securing device, however, is some form of bent-up sheet-metal tab that bears against a flat on the nut or bolt head. Sometimes the tabs will be part of an arched plate that wraps across the bottom of the rod, or it may be a part of the oil-dipper assembly. Usually a little prudent prying and tapping with a screwdriver will bend the tabs out enough for the fastener to be loos-

ened. If the tabs do not appear to have been removed before and are in good shape, you can probably use them again. But tabs and lock plates that show signs of previous work can break, introducing a piece of stray metal into the gears or other parts or allowing a connecting rod cap to loosen with disastrous results. They're inexpensive and you'll feel better using new ones.

Piston-Pin Removal

Most piston pins in small engines are retained by spring-wire snap rings or similar retainers that seat in shallow grooves in the piston pin bosses, preventing the pins from sliding out endways and damaging the cylinder bores. Such retainers can usually be sprung out of the grooves with a pair of needlenosed pliers, but as you do this, check them for security.

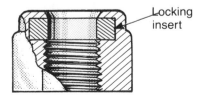

Locking insert

This self-locking nut uses a plastic insert collar to grip the threads.

Many connecting-rod bolts are secured by lock tabs that bend up against the bolt heads (above). A light tap with a punch, below, gives access to the bolts, which are then removed in the usual manner.

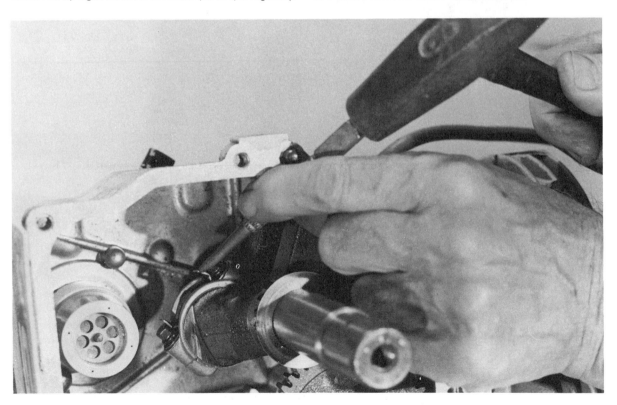

If you detect looseness in the ring or if it seems to slip out too easily, the groove in the piston is suspect. If the groove appears to be battered or worn, the piston must be replaced.

Aside from the preceding point, you probably don't know if you'll need a new piston or not. For that reason, work carefully to avoid damaging what may be a perfectly useable piston.

1. After removing both retainers, you may be able to push the pin out with your thumb or a piece of wooden dowel. If the pin falls out or comes out too easily, the pin and piston may be worn.

2. Piston pins are more likely fitted to what is termed a *hand-push* fit. Moderate-to-firm palm pressure against the pin should push it through the bore. If it doesn't push out readily, check for gum deposits just outboard of the pin-retainer groove. Also look for brownish gum or varnish deposits on the pin at each side of the connecting-rod bearing. If found, clean them out with carburetor cleaner and try again using palm pressure.

3. If the pin still won't budge, try fitting a hardwood dowel against one end, hold the opposite side in the palm of your hand, and tap the dowel with a hammer. *Do not place the piston against a hard surface when tapping or you may distort it.*

4. If tapping doesn't work, try heating the piston in boiling water or hot engine oil until thoroughly hot—making sure the piston does not rest directly on the bottom of the heating vessel. This type of fit is common in some engines, and expansion will usually free the pin so it may be pushed or tapped out readily.

A really tight pin that resists the preceding measures will require some type of puller. The trick is to remove the sticky pin without marring or distorting the piston. Not all piston pins are hollow but most are. You can make a puller from a hardwood block curved to fit the piston outer diameter and bored to allow the pin to pass through. With this you can insert a long bolt through the pin center with a washer and nut at the other end; the washer will bear against the pin. On the block side use a piece of pipe or tubing that will allow the pin to pass through, and top it with a metal plate and washer. Tightening the nut will pull the pin.

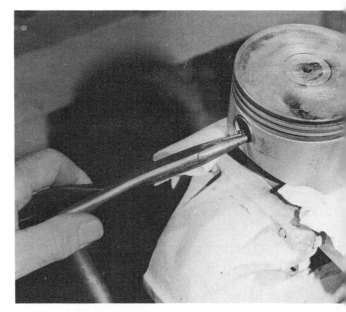

Snap rings that retain the piston pin should take a little effort to remove. If they slip out easily, they're distorted or the groove in the piston is worn.

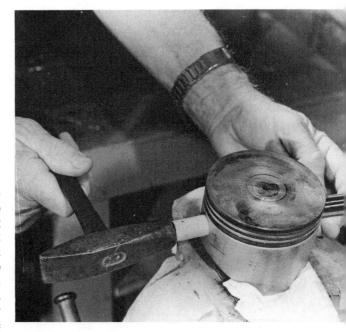

Tap out sticky piston pins with a wooden dowel. If the pin doesn't come easily, heat the piston in boiling water or hot oil and try again while it's hot.

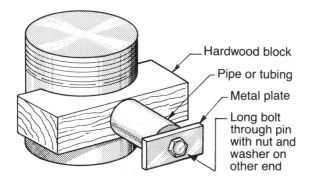

Hardwood block
Pipe or tubing
Metal plate
Long bolt through pin with nut and washer on other end

If a piston pin is really tight, make a puller of hardwood with a hole somewhat larger than the pin. Again, hot water or oil should help.

Piston Pin-Bore Inspection

After removing the retainers and pin, a general visual inspection will give you a hint or two as to whether a piston is likely to be fit for further use. As I'll explain, this is one of those engine-rebuilding situations where one thing can lead to another and there may be several options.

Check the condition of the piston pin and the pin bores first. If you find the pin is out of round or that the diameter is below specifications, it may be that the wear is confined to the pin but don't bet on it. You could buy a new pin and find that the pin bores were also worn oversize, or that the bearing in the upper end of the connecting rod was also worn and loose. This bearing may simply be in the metal of the connecting rod or there may be a pressed-in bushing. Here is where decisions and options enter the picture.

If you can buy an oversize piston pin, perhaps with .005 inch extra diameter, you could have the piston-pin bores over-sized on a precision hone. But then you also must have the connecting rod upper bearing resized. Both of those jobs are routine at automotive specialty repair shops. Going that route may prove economical, provided the main piston dimensions and ring grooves are satisfactory on inspection. But if the basic piston is marginal, a new piston and pin is definitely a better choice.

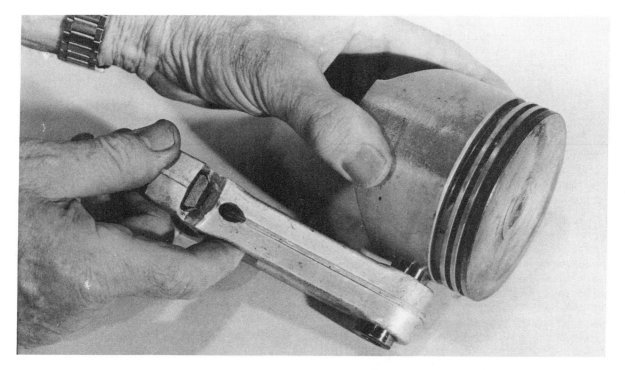

If you can't produce more than the slightest wobble with your fingers, piston-pin fit is probably satisfactory.

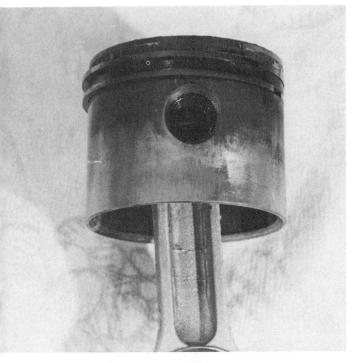

This piston has been running under stressful conditions. Carboned rings are stuck in places, and the darkened area indicates combustion gases have been passing.

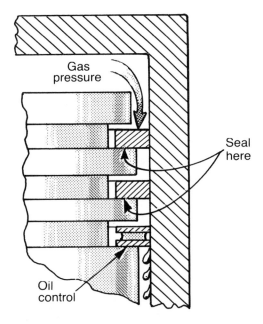

The force that pushes rings against cylinder walls comes primarily from gas pressure behind the rings.

As for re-using the connecting-rod upper bearing, your choice here will be governed by the condition of the crankpin end of the rod. If it has replaceable bearings, you may be able to use it. If the bearing is simply part of the rod metal and shows distress or wear, you'll need a new rod, though you wouldn't want to oversize the upper end of a new rod to salvage an old piston. Again, this is a typical rebuilding situation where trade-offs and judgment play a large part.

Piston Inspection

A piston that is clearly battered on the crown from a loose object in the combustion chamber, or has burned notches around the top edge—often called the "fire land"—is not a candidate for re-use. But if the piston does not show major scratches or

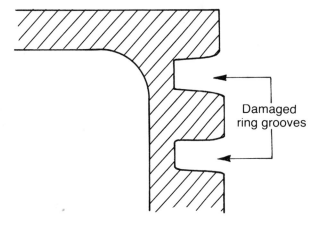

Some common practices for cleaning carbon and gum from ring grooves can result in grooves that are flared out or rounded off at the corners.

score marks, and its pin still fits snugly, it may be quite useable. The only way you can tell is to check dimensionally against the manufacturer's wear limits. Before you can do this, you must remove the old rings, which never can be reused, and thoroughly clean the piston and ring grooves.

Removing piston rings. Remember that the rings must not only seal the combustion and compression strokes against the escape of air and gases, they must also control the oil on the cylinder walls. Effective sealing results from gas pressure that enters behind the rings and forces the rings out against the cylinder wall. If the rings fail to seat against the lands, gas pressure is lost and excess oil can pass by. Thus, the seal between the sides of the rings and the groove lands is critical.

Assuming your piston rings are not stuck solidly in the grooves, a ring expander or light pressure from opposing thumbs should spread them and lift them free. If you have a piston with the rings solidly embedded in carbon and hard deposits, the piston should be discarded.

Cleaning ring grooves. Some mechanics run a file tip or a screwdriver around the groove to remove carbon. Others will use a piece of abrasive cloth, folded and run back and forth in the groove like a shoe-polish cloth. The first method nicks and scrapes the very faces necessary for good sealing. The second one flares out the grooves and generates a shape that will never seal a ring.

The best way to clean a piston is to chuck it carefully in a lathe and use a tool slightly narrower

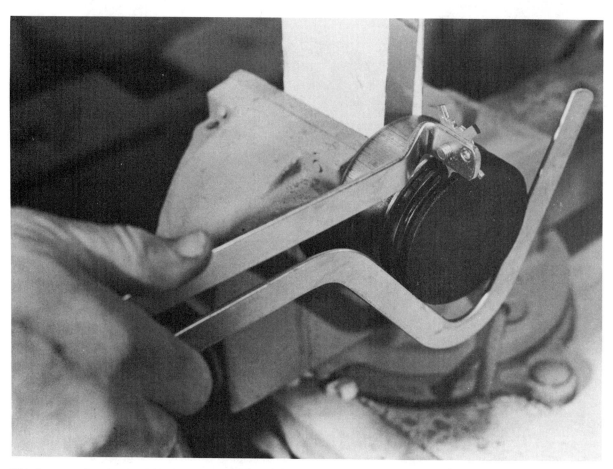

This inexpensive ring-groove cleaning tool has several cutter widths to fit various pistons.

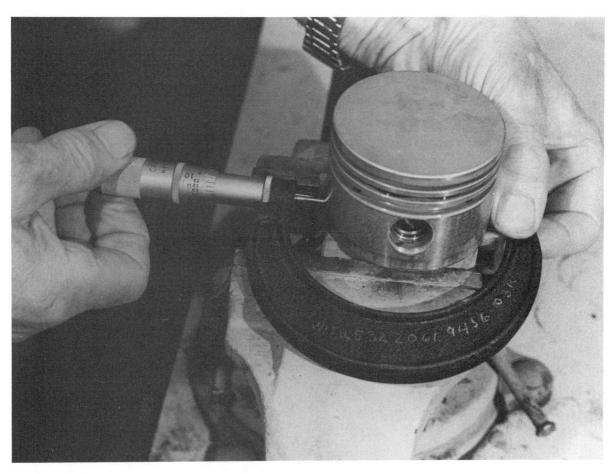

Make a micrometer check of the piston diameter at the bottom of the skirt and just below the ring grooves. Then check the manufacturer's specs for your engine to see if wear is serious.

than the groove width. The corners of the tool should not be sharp since you do not want a sharp corner at the back of the groove. Turn the lathe by hand and slowly feed the tool in to clean the carbon from the groove. A small metalworking lathe is ideal.

The second way to clean ring grooves requires a plier or tong-like tool with one leg that aligns in the groove and a cutting tooth on the opposite side. This tool is worked by hand with the piston flat on the bench or with the rod in place and clamped between padded vise jaws.

Once the rings are off and ring grooves cleaned,

make the first check of the piston with micrometers at the bottom of the piston skirt and at 90 degrees to the pin bores. Repeat in the same area but in line with the pin bores. This portion of the piston receives the least wear and should closely approximate the manufacturer's dimensions. Wear here suggests the engine ran with very dirty oil or without a functioning air cleaner.

At this point you may get a surprise and find out that your piston is not perfectly round but is slightly oval. Generally, that means the builder is using a *cam-ground piston* to compensate for uneven expansion when the piston is hot and make for a quieter

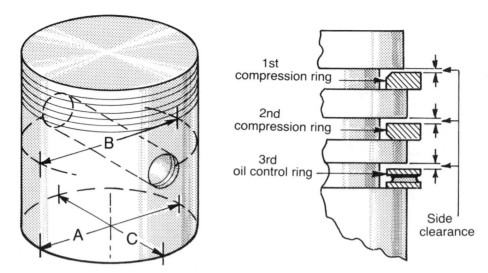

Piston clearance between lower skirt bore should always be measured at 90 degrees to the piston pin, dimension *A*. On a cam-ground piston this will differ from *C,* and if it's taper-ground *B* is slightly less than *A*.

running engine. If all sections of the piston wall were of the same thickness, the heated piston would probably expand more or less evenly; but heavy sections of metal are needed for the pin bosses, making the piston expand unevenly. By starting out with a slightly oval shape, the piston actually approaches round when hot and can be fitted more closely. Not all small engines have that feature, and often you can hear a definite piston slap as the piston oscillates in the cylinder from combustion forces.

Repeat the above measurements above the piston bores and just under the bottom ring groove. Again, you may find that the piston has some unsuspected features. If the upper measurements are slightly less than at the bottom of the skirt, the piston is also *taper-ground.* That compensates for the additional heat at the upper end. If you are now satisfied that your piston is in acceptable shape with respect to the major diameters, it's time for the next, critical phase of piston inspection.

Ring grooves and lands. You'll need a new set of rings against which to measure groove wear.

Most small engines have a top or compression ring, a second compression ring, and a third, oil-control ring. The oil ring will usually be slotted or vented in some manner to trap oil and allow it to escape. A top compression ring can often be identified by a chamfer on one inner edge. The chamfer imparts a slight twist to the ring and is normally installed towards the top of the piston. Sometimes rings are marked "top" to show which side goes up. The second ring may also have a chamfer or notch on the inner edge, but this notch usually faces down. It is very important to orient the rings according to the manufacturer's specifications or the directions on the ring package.

If you are rebuilding your engine, you will almost never use so-called production rings unless you have honed the cylinder oversize for a new piston. Service rings for worn cylinder bores are normally required for a simple ring replacement without honing or reboring the cylinder. Such rings are often faced with a special coating and are designed to seat in cylinder bores that are within limits but less than factory-new and lack the sharp, crosshatch honing marks found in new cylinders.

In addition, different ring material and facing surfaces are specified for different cylinder bores. A ring set intended for an aluminum cylinder is not suited to a cast iron bore, and vice versa. Thus, be sure to check the parts listing and manual carefully before buying a new set of rings.

Piston rings must be checked for fit in two ways: ring gap, and ring side-clearance in the grooves. The check for ring side-clearance will show up ring grooves worn in a flared-out pattern. In most cases, side clearance is best checked by slipping the ring into the groove from the outside and more or less rolling it around with the feeler gage alongside. Remember, the rings are standard; what you're checking is the condition of the piston.

The second check, for ring gap, is made by using the piston to push the ring down squarely into the cylinder bore towards the lower end, which will be the tightest point and where the gap will be smallest. Withdraw the piston and measure the gap between the ends of the ring with feeler gauges. In no case should the ring ends butt together and the gap should always be at least that specified.

While it was once routine practice to file the ends of a ring that had too small a gap, that is seldom necessary with service rings from the manufacturer. If the bore is right, the gap will be right. The most likely possibility if you find a set of rings with tight gap is that you have somehow obtained a set of oversize rings for a resized cylinder. *Do not attempt to use them.*

Above, side clearance is measured by rolling the ring around the groove and making feeler checks as you go. Feeler gauge is also used to measure ring gap (below). Push the ring in with the piston so it's square in the bore.

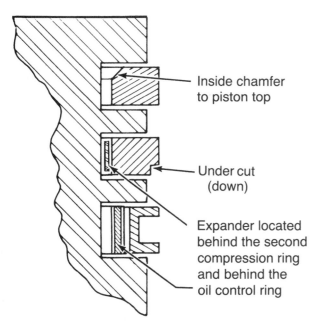

Inside chamfer to piston top

Under cut (down)

Expander located behind the second compression ring and behind the oil control ring

Service rings for used pistons and cylinders are seldom identical to production rings. Among other features, both lower rings have a spring expander behind them for better wall contact.

REINSTALLING RINGS AND PISTON

Piston rings for small engines are relatively light and easy to expand. They are also brittle and easy to break if you expand them a hair too much. The only way to install them safely is with a ring expander adjusted to open them only barely as much as needed to slip them over the piston. Trying to pop them apart with your fingers or thumbs—as you did when removing the old rings—is asking for breakage.

Never try to put rings in the grooves by spiraling them over the piston and through the upper grooves. Also, in many two-stroke engines you'll find small pins in the grooves to prevent the rings from rotating. This avoids having the ring gap align with one of the ports where the ends of the rings could catch. When installing rings in such a piston be very careful to align the ring gaps and locator pins.

After placing the rings in the grooves of a four-stroke piston it is customary to slide them around so the gaps do not align and create a direct channel for hot gases to pass through. On a three-ring piston, locate the gaps about 120 degrees apart. It is not necessary to be exact as long as the gaps are in random locations relative to each other. Actually, when the engine runs, piston rings tend to migrate around the piston freely. Tecumseh advises against placing a ring gap towards the valve side of the cylinder. This relates to a depressed or "trenched" area in the block where the rings could catch when

This ring-expanding tool holds the ring ends in line and allows you to squeeze gently, as shown, until the ring is properly installed. Using your fingers or other crude methods only invites trouble.

On two-stroke engines you'll usually find a tiny pin that prevents the ring from rotating in the groove, and the ring ends from aligning with a port. Be aware of this pin's location when you install new rings.

they slip free of the ring compressor on installation; it is not a factor once the piston is installed.

Before installing the piston, rings, and connecting rod assembly, make a final check to be absolutely certain that no grit or dirt from the workbench is adhering to the piston or rings. At this point, the rings are expanded out beyond the surface of the piston by their normal tension. They must be compressed flush with the piston before they will slide in, requiring some type of ring compressor.

In production, it is common to use a tapered *ring pot*—a ring of hardened steel with a tapered inner bore that gradually squeezes the rings as you push the piston through. Such tools are nice if you are installing pistons hour after hour, but they're impractical for home use since you need a precise size for each engine with a different bore size. I mention the tool because you may find the same principle built into some two-stroke engines where a conventional compressor can't be used. If you happen to work on such an engine, you'll recognize it by the tapered area at the bottom of the cylinder.

A practical alternative is a ring compressor that squeezes all the rings flush with the piston surface and allows the rings to slide into the cylinder without catching. A ring compressor that will fit all common sizes of small engines is not expensive and will last a lifetime. It typically consists of a spring-steel sleeve made up by wrapping thin metal around on itself so it can be squeezed down and still remain round. A band with a ratchet-like drive surrounds the sleeve, and by tightening the band the sleeve is compressed and the rings forced into the grooves.

A word of caution is in order about trying to install rings with crude techniques. Piston rings are delicate and the faces are easily nicked or damaged.

By compressing the rings snugly inside this spring-steel compressor, you can easily slip the piston into place in the bore.

Never attempt to enter the skirt of the piston into the cylinder bore and then poke in the rings with a screwdriver or like tool while wiggling the rings into the bore. This is certain to cause damage.

Installing Four-Stroke Pistons

To use a ring compressor, place the well-oiled piston inside and carefully tighten the sleeve. If the piston has locating pins for the ring gaps, be certain the gaps are positioned at the pins since tightening with the gaps out of place will definitely break the rings.

Tighten the compressor sleeve firmly and then back off a click or two on the ratchet release so you'll be able to slide the piston out of the sleeve and into the cylinder. The usual degree of tightness is enough that it takes a light tap with a soft mallet or hammer handle against the piston crown to move the piston. Try to work very evenly to avoid cocking the piston in the bore or allowing a ring to pop out and catch on one side.

Another caution is in order here. Though you'll probably be concentrating your attention on the pis-

Tapered area

Some two-stroke engines have a tapered section at the bottom of the bore to compress the rings.

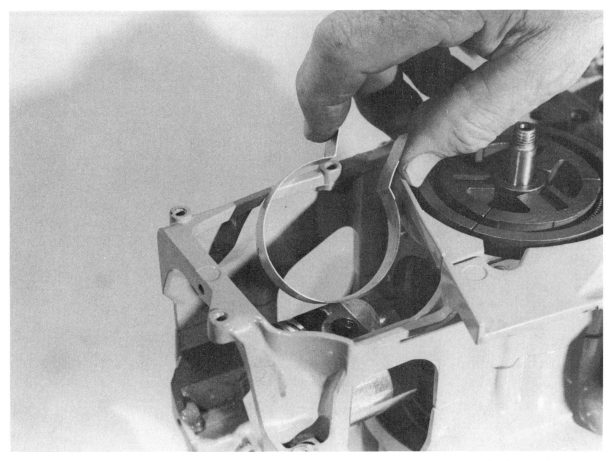

Make a ring compressor for small two-stroke engines from a strip of aluminum bent to the piston size.

ton and rings, don't ignore the lower end of the connecting rod. There will usually be two studs or bolts projecting downwards, which will later retain the rod cap. Engine manufacturers in production often cover these studs with short lengths of rubber hose or plastic sleeves so they won't hit and scar the crankpin when the connecting rod projects down into the crankcase.

In your small engine all that's probably necessary is to make sure the crank is turned so the crankpin is at bottom center. At the same time, observe and guide the lower end of the rod as you tap in the piston. This is also the time for a final check of the piston and rod markings and their proper orientation. It is quite possible to become so interested in getting the rings in that you put the piston in wrong.

Installing Two-Stroke Pistons

Installing a piston and rings is easier on a fairly large four-stroke engine than on a small two-stroke. In the latter case, you are attempting to place the cylinder *over* the piston and rings rather that placing the piston *into* the cylinder. That means you'll be working in a cramped area between the crankcase and the bottom of the cylinder, and the connecting rod will want to flop back and forth on the crankpin. You may find it difficult to keep the ring gaps aligned with the locating pins, if pins are used. Also, the rod is fairly light and can easily be bent or twisted.

You might try making up a simple, band-type compressor. You can't use the sleeve compressor

With the crankshaft in the position shown, the rod bolt could put a nasty nick in the highly polished crankpin.

since it would encircle the rod and you couldn't remove it.

1. Cut a strip of roughly .30-inch aluminum, slightly wider than the width of the ring band.
2. Form the strip into a circle with two bent-out tabs projecting. Make sure the diameter of the circle matches, or is slightly less than, the diameter of the piston.
3. By squeezing with your fingers or with small locking-type pliers, you can grip the tabs and compress the rings.

A still different two-stroke engine design requires a two-handed approach with someone to help you —at least until you've done it a few times. Here, the crankcase is split parallel to the crankshaft and the open halves of the upper crankshaft bearings are

facing up towards you. The crankshaft, connecting rod, and piston are preassembled and torqued on the bench, and the entire assembly is then lowered to install the piston and rings and seat the crankshaft in the bearing halves at the same time. Enlist someone to support the crankshaft and suspended parts while you align the piston and compress the rings into the lower end of the cylinder bore. Lacking help, it may be easier to position the cylinder and crankcase flat on the bench and work slowly by sliding the assembly gently sideways.

Torquing Connecting Rods

Correct tension or torque on the rod bolts or nuts is one of the most important factors in a successful engine assembly. Very few modern engines, except those with needle bearings, have replaceable lower connecting-rod bearings. If the bore for the crankpin or piston pin is damaged or worn beyond limits, the entire rod is replaced. Sizing of the bore is done when the rod is manufactured. Thus, if you torque the rod fasteners correctly you'll have a proper fit and oil clearance.

1. Before installing the rod cap, make a final check to be sure the crankpin is perfectly clean. Then coat the crankpin with clean oil.
2. Also be sure the oil hole and alignment marks are correct. Note, for example, that connecting rods on vertical-shaft engines have oil holes located differently in the rod flanks than do horizontal-shaft connecting rods. *Take nothing for granted at this critical assembly point.*
3. Turn the fasteners down by hand to be sure the threads are engaged properly and not crossed. Many self-locking fasteners resist turning after the first few threads because their locking feature comes into play. Continue turning down the fastener until it snugs up against the rod shoulder or dip-oiler device, then give it an initial light lightening with the wrench.
4. From here on, work from side to side with the torque wrench in moderate increments until final torque is reached.

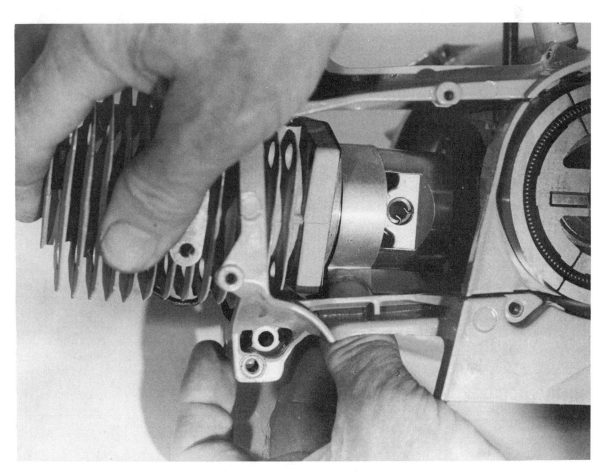

Align the ring gaps on the pins, squeeze gently, then work cylinder down over piston. Don't use force.

If a light tap or firm finger pressure on the crankpin can just move it a few thousandths of an inch either way, the fit is about right.

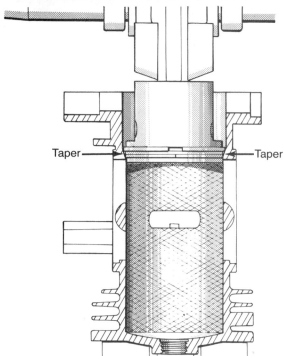

Taper — — Taper

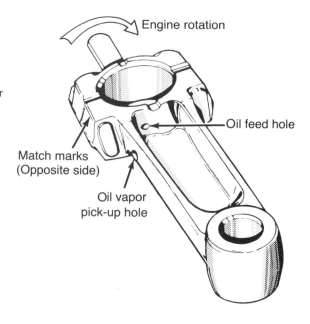

Engine rotation

Oil feed hole

Match marks
(Opposite side)

Oil vapor
pick-up hole

On some engines the crankshaft, connecting rod, and piston are bench-assembled and lowered as an assembly into upper half of crankcase and cylinder.

This rod has an oil-pickup opening and a drilled passage to the rod bearing. Oil collected in the rod channel feeds into the bearing by centrifugal action.

Connecting-rod fit. After installing the rod-bearing cap, check for excessive snugness or looseness. If the rod is perceptibly loose on the shaft, either the crankpin is worn below limits or the rod cap is improperly installed. An example would be a rod that was split by fracturing and on which you have reversed the cap position. It is also possible that you have a crankpin that was ground undersize at some time.

There are several ways, including micrometer checks, to determine if your perception of looseness is correct. The easiest method is to remove the rod cap, clean off the oil, and place a piece of Plastigage along the bearing lengthwise to the crankpin. Reinstall the rod cap to the proper torque. Plastigage is a semisoft, wax-like material that's sold in automotive repair shops and resembles small wires of plastic or wax. When squeezed down in the bearing it will spread out to a width dependent on how tightly it's squeezed. The width of the Plastigage material left after removing the bearing cap again is translated into bearing clearance by matching against a series of markings on the edge of the envelope in which it comes.

Be sure the surfaces are clean and dry when making a Plastigage check since oil causes the material to smear. If, after the check, you determine that you have a mismatched crankpin bearing, you will have to measure both the crankpin and the rod bore to determine which is at fault.

A connecting-rod bearing normally fits on a crankpin just snug enough to require firm finger pressure or a light tap with a wooden drift and light hammer to slide the rod slightly along the crankpin in either direction. If you have such a fit and the engine turns over by hand freely and normally, you've got a good assembly. Obviously, the preceding remarks apply to *conventional* and not *needle* bearings.

FITTING BEARINGS ON ANTIQUE ENGINES

Nearly all of the instructions in this book relate to modern engines. Nevertheless, there is widespread interest in rebuilding antique and collectible engines, mostly for fun. Such engines can be put back into running shape even though parts aren't available and, perhaps, never were. Some of the old connect-

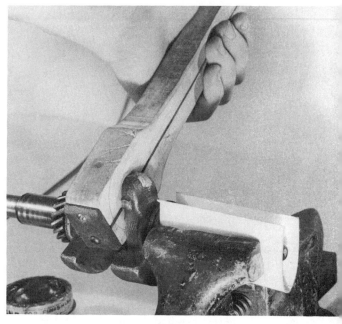

A wooden lap and fine valve grinding compound are used to clean up an old crankpin. Work slowly and measure often.

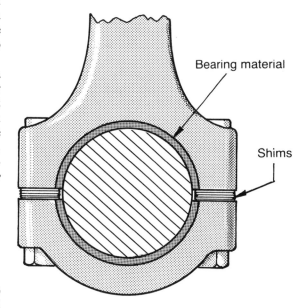

The rod-bearing material may also need cleaning up, in which case honing is necessary after removing some shims. If the clearance is right vertically, a little extra at the sides won't hurt.

ing-rod bearings, for example, were babbitt metal melted and spun into place in the rod bore where it permanently adhered. After the metal was poured into the bearing it was machined and bored to fit whatever size the crankpin was. Thus, in truth, there never was a connecting rod bearing as a separate part for that engine.

You may find that such bearings have a thin pack of brass shims on each side between the cap and the rod. Those allowed you to "take up" a noisy rod by pulling out one or two shims. If such a bearing is merely scored lightly, but not melted or pounded out, it can be resized. If it is merely worn, you are in luck. Most of the wear occurs at the top and bottom of the bore and very little at the sides. By closing up the spacing between the cap and rod faces you will probably have plenty of metal to resize.

First, though, check the crankpin for wear, out of round, and taper. Many of the old crankpins were not induction hardened and were quite soft. You could take such a crankshaft to a regrinding shop if you find one that can handle the job—and it can be costly. Or you might try hand-lapping the crankpin back into shape.

A wooden lap, bored slightly under the crankpin diameter and hinged with belt leather, will do the trick. Use valve-grinding compound and check what's happening frequently. With a little care you can probably restore the crankpin to almost perfect roundness and straightness. It will, of course, be undersized by the amount of metal you removed.

This means that you cannot ream, hone, or bore the bearing undersized to fit the crankpin without somehow first reducing the bore diameter of the existing rod to get some metal to work with. You must reduce the diameter to slightly less than the crankpin diameter. Sometimes, just the removal of a few shims is sufficient. Other times you will have to remove metal from the mating face of the rod cap. The best way to do that is with a flat plate or piece of glass and lapping compound.

1. Place the rod cap on the plate or glass in a puddle of lapping compound and work it in a figure-eight motion.
2. Periodically clean the cap and try it on the pin for fit.
3. When you've removed enough metal, there will be a gap of .004–.005 inch between the cap and rod on each side.

When you bolt the cap to the rod the bore will be other than round, with the vertical dimension somewhat smaller than the horizontal dimension across the split line. The best way to handle this is to take the rod to an automotive repair shop and have them hone the bore to the crankpin diameter. If you have a suitable adjustable reamer you can do the job yourself. Afterwards, with some Plastigage and one or two shims about .0015 inch thick, you can come up with a nice fit. If you lack brass-shim stock, try aluminum-foil kitchen wrap.

That method of repair usually works out because the babbitt-metal bearing material is forgiving and will shape itself to fit a slightly imperfect crankpin. Millions of automobile and other engines have been fitted this way. On the other hand, modern rods with the bore directly in the aluminum and a hardened crankpin may not be conducive to such a repair.

18

The Cylinder Bore

Whether you tell your local parts agency you want a cylinder block or a crankcase for a given engine, in most cases you'll get the same part. That's because most small four-stroke engines consist of a single basic structure with the cylinder and crankcase cast as a single piece. That is not often true of a chainsaw or other small two-stroke engine because the crankcase is usually part of the saw structure or other member.

WHAT THE CYLINDER DOES

The cylinder must be strong enough to resist breakage or serious distortion under high combustion pressures and temperatures. Since those temperatures may reach 1,600°F under load, the cylinder walls must be cooled by the fresh air/fuel intake charge, splashed oil, and air flow through the fins. That's why the air intake screen and cooling fins must be kept free of grass and cuttings. If the oil temperature is allowed to get too high, the result can be a scored piston and cylinder.

You'll also find different wear patterns in the cylinder bore, which reflect the difference in temperature between the upper and lower ends. The top end of the bore will always show greatest wear because it is directly exposed to flame and receives less oil. If you have the cylinder head off and run your thumbnail up at the top edge of the ring travel you'll often find a wear ridge or ring step, even though the bottom of the bore is not worn at all.

Cylinder-Bore Materials

There's no more critical running surface in your small engine than the cylinder bore. Cast iron has always been the material of choice. Iron has a relatively low expansion characteristic and remains quite stable dimensionally even under extremes of heat. Equally important, cast iron has a peculiar grain structure that includes free carbon. The carbon, together with the surface characteristics of honed iron, makes it an ideal running surface. Once broken in, a cast-iron cylinder will run for many years and eventually acquire a surface mechanics refer to as a *glaze*.

Aluminum cylinders. Because cast iron is an ideal cylinder bore material, most heavy-duty small engines with light aluminum structures and fins will have a cast-iron sleeve or cylinder liner permanently cast in place. Lighter-duty engines such as those used on mowers, however, often use the aluminum itself as a bore material. Though we may think of aluminum as a soft metal, it has excellent wearing properties and will reliably serve in such an engine for many years.

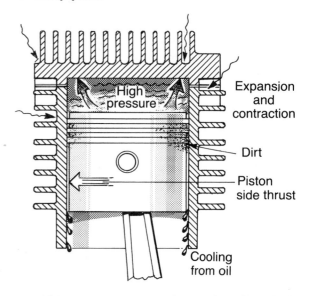

In addition to other stresses imposed on the cylinder bore, the oil introduces a thermal strain by reducing the temperature at the lower end of the cylinder.

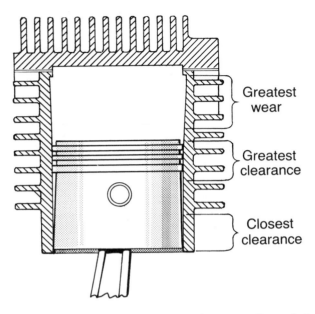

Wear will be greatest in the upper bore near the end of the piston-ring travel, while practically no wear occurs at the bottom of the bore.

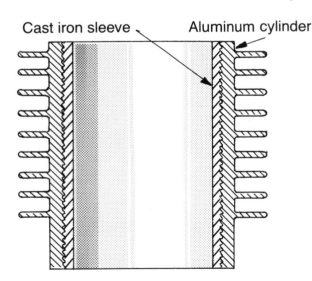

The outer surface of a cylinder liner is usually machined in a grooved or rough pattern so it is well-secured and can conduct heat efficiently to the outer aluminum walls and fins.

BREAKING IN A CYLINDER BORE

Engine manufacturers control the final finish of the cylinder to something less than the smoothest surface possible. The glass-like finish, or glaze, that comes after long operation is fine for running seated rings; but it will probably prevent new rings from seating, thus causing excessive oil consumption. For that reason, many shop manuals over the years have recommended "breaking the glaze" as a minimal resurfacing procedure. Often this amounted to no more than using fine abrasive paper and kerosene to scour the cylinder bore until dull.

Today, your best bet is to check a manual for your engine or the instructions that come with the replacement rings. Aluminum cylinder bores, for example, are generally not treated to glaze removal. Moreover, many replacement rings for cast-iron cylinders have chrome or other special facings and will seat without glaze removal. *It is very important that new rings seat in properly, so follow any instructions carefully.*

Honing

The honing process to final size and finish is the basis of a cylinder that will have good compression and minimum oil consumption. If you followed the path of a single abrasive grain in the hone as it both rotated and moved lengthwise in the bore, you'd find that the two motions combined to produce a spiral path. Overlap thousands of such paths and spirals and you produce a crosshatch pattern. Such a pattern actually consists of millions of little peaks and valleys. During break-in the peaks gradually wear away until an equilibrium is reached and further wear becomes minimal. Some wear also occurs on the piston rings, which are now seated. The engine has put its own finish on the cylinder bore.

After this break-in period specified for your engine, it is important that you drain and discard the oil if the engine is a four-stroke. It contains myriads of minute metallic particles. Do this according to the break-in operating schedule and time recommended by the manufacturer.

Throw-away engines. Not all small engines are expected to be rebuilt. Providing a stock of oversize parts would add to the initial cost of the engine, and often the cost of the machine is so low that the labor costs of rebuilding after several years would be higher than the cost of the tool or machine itself.

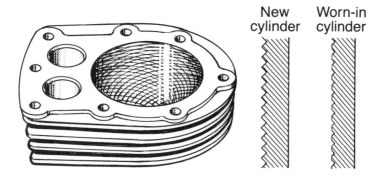

New cylinder   Worn-in cylinder

Glass-like glaze that develops on a worn-in cylinder, depicted above, can be removed if required with kerosene or light oil and medium-grit abrasive paper (below). Be sure this is recommended for your engine.

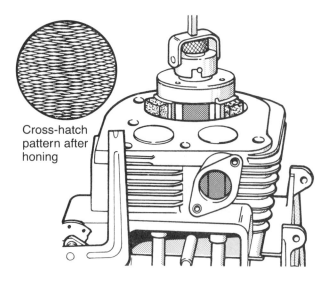

Cross-hatch
pattern after
honing

Cylinder hone, shown entering the bore, carries several abrasive stones that are lightly tensioned to press against the walls.

Very small two-stroke engines often fall into this class, as do some other engines used in light-duty lawn and garden equipment.

CYLINDER-BORE INSPECTION

If you plan to rebuild your engine, your first step is a thorough inspection of the cylinder bore. To do this properly, you must disassemble the engine and remove the piston and probably the crankshaft and camshaft for good access, as shown on pages 238 and 245. Before going further, wash the cylinder block with solvent and dry it; working with dirty, oily parts is never good practice.

If your preliminary removal of the cylinder head showed a definite ring step or wear ridge at the top of the piston travel, remove this ridge before attempting to force the piston out past it. In all cases it must be removed before installing new piston rings. That is because the top ring, as it wore into the cylinder wall, assumed a rounded, worn profile on the upper edge that matched the ring step in the wall. A new top ring will have a sharp edge that would impact on the step left by the old ring. The result would be a broken ring or damaged piston, and the engine would have a sharp knocking sound.

Ideally, you should remove the ridge with a tool known as a *ridge reamer*. The tool centers in the cylinder bore and has a sharp cutting edge that projects slightly from the rotating center-part of the tool. The latter is expandable, and by taking successive light cuts by hand and expanding slightly after each cut, the ridge is soon cut back flush with the rest of the cylinder wall. Sometimes the ridge is minimal and can be removed by hand-scraping and working a small hone around the top of the cylinder.

Visual Inspection

Make a preliminary visual inspection of the cylinder block for cracks, scores, damaged head-bolt threads, raised areas in the head gasket area, damaged cooling fins or other damage such as cracks in the valve seat area. Use a good flashlight or work light so you can see clearly down into the bore.

Different engine manufacturers have different wear limits for their engines. Two important dimensions are usually listed in a manual: *oversize wear limit* and *out-of-round.* The out-of-round condition develops sometimes because of heat distortion, but mostly because of piston side-thrust from the mechanical forces of combustion pressure and the crankshaft angle. As the piston comes down on the power stroke, it is forced sideways against the cylinder wall. A typical out-of-round limit for a small engine might be .0015 inch, while a typical wear limit or oversize is .003 inch. If either the out-of-round or the oversize exceeds the limits for your engine, you must hone the cylinder bore to the next oversize for an oversize piston.

The manufacturer's wear limits aren't necessarily carved in stone. The maker is telling you that if you replace rings without correcting the bore conditions, the rings probably won't seat in quite as well, you can expect some higher oil consumptions, and the engine will likely be a trifle low on power. Often on an occasional-use small engine, those conditions can often be tolerated if you're willing to settle for less than like-new peak performance.

Measuring The Cylinder Bore

Cylinder-bore measurement requires some form of inside micrometer. Start by setting the micrometer to the standard bore dimension for your engine.

Bright ring at the top of the cylinder bore above shows where the ring step has been cut away. If the step is not removed, the new ring's sharp edge will hit it and possibly break the ring or damage the piston.

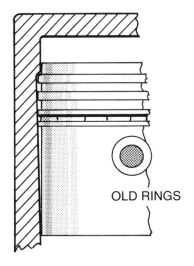

OLD RINGS

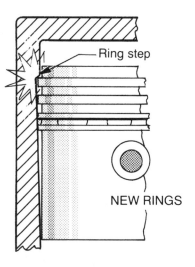

Ring step

NEW RINGS

Before replacing a cylinder head, run a flat file or hone over the gasket surface.

1. Pass the gauge into the bore in a plane parallel to the crankshaft. You'll have to rock the micrometer slightly until you feel the contact drag against the walls.

2. Try the standard dimension first down at the bottom of the cylinder. Wear is normally minimal there, and you may even find that the bore is still standard size.

3. Gradually work your way up through the cylinder, taking measurements as you go. As you reach a point about midway up in the piston-ring travel you'll find that the bore diameter increases perceptibly and the cylinder bore has actually worn in a taper. *Remember, these readings are all taken parallel to the crankshaft—the direction of least wear.* Jot down the readings as you make them.

4. Next, repeat the inspection at right angles to the crankshaft. Don't be surprised if the wear pattern is substantially higher. *This is normal.* You'll probably find the greatest wear in the upper-ring travel area at right angles to the crankshaft.

Usually, if you take readings at six points up and down the bore you'll acquire an accurate picture of its condition. Once you know the cylinder-bore dimensions, you are in a position to decide whether the cylinder must be honed oversize for a new piston or if simple ring replacement will suffice.

OVERSIZING THE CYLINDER BORE

All that follows assumes that the economics of resizing a cylinder bore for a new piston are favored

An inside micrometer is the usual tool for measuring bore wear. Start at the bottom of the cylinder, then take other readings first midway up and then at the top of the ring travel.

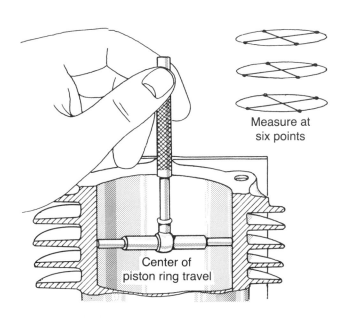

Measure at six points

Center of piston ring travel

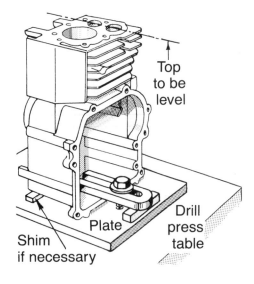

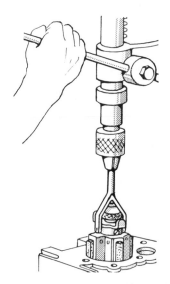

A home-shop drill press works fine for honing a worn cylinder. The plate is not clamped to the table but is lubricated and free to slide so the hone self-aligns with the bore.

over a new short block. If the cylinder bore is scored or is significantly oversize and out-of-round, it must be honed to a standard oversize, usually .010 or .020 inch over. Most small-engine builders offer hones for this purpose. Again, you must know the proper finished dimensions so the new piston will fit correctly. One point not often discussed is that engine builders may occasionally use an oversize bore and piston in a new engine. That is done to salvage a cylinder that may have been accidentally oversized in production or that had a casting flaw that wasn't quite cleaned up at the standard bore dimension. Such blocks will usually be stamped to indicate the oversize to which they were built.

The Honing Setup

Some form of power drive is required to turn the hone in the cylinder bore. Although a slow-speed ³⁄₈- or ½-inch electric drill may be used, it makes controlling the desired vertical travel speed difficult. A drill press is much easier to use. If the process attracts you, a fine honing job is possible in the home shop if you can borrow or rent the honing equipment.

In use, the hones are mounted in a stone carrier

that may be expanded or contracted to cover a limited range of bore sizes. That means you must have the correct size hone carrier for your engine. A drive extension is chucked in the drill press and engages the stone carrier in a universal-type joint so that the carrier is free to align with the cylinder bore. The hones must be able to align and center, so this freedom is essential.

Also make sure the cylinder is held in vertical alignment with the hones and spindle, and that it is free to slide on the drill-press table and make minute self-adjustments with the hones. The easiest way to do that is with a steel plate about ½-inch thick, which has either clamps or bolt holes to mount securely to the base of the crankcase. The heavy plate prevents the crankcase from raising up when the hone is raised, yet still permits it to slide for alignment.

With a horizontal-shaft engine, if the crankcase is flat on the bottom, all that's necessary is to clamp it to the plate after cleaning up any rough spots or burrs with a file. Use shims if needed to square the bore of the cylinder with the drill-press spindle. With a vertical-shaft engine, it will be necessary to use a backing plate bolted or welded at a right angle to the base plate.

Be sure the drill-press table is flat and smooth and oil it so the plate will slide. Adjust the drill-press travel so the hones will project somewhat—about ½ to ¾ inch at the bottom of the bore when the drill-press lever is at full down travel. A similar projection should be provided at the top. These top and bottom projections permit the stones to wear evenly and cover the entire bore area. In some cases it may be easier to use a block of wood to limit the lower hone travel.

It is also a good idea to provide some sort of catch pan under the drill press table since you will be using liberal amounts of honing lubricant. Different hones require different types of lubricant, and you should use whatever type is specified by the hone maker. Set the drill press for 350 to 700 rpm. On most home-shop drill presses this is the low-speed pulley ratio, commonly about 500 rpm.

Honing technique. A certain amount of "feel" and listening helps when honing a cylinder. If you set the hones to ride too hard on the cylinder wall, the drag will cause overheating and wear the stones prematurely. Too light a cut wastes time and doesn't give the correct finish. Slowing or speeding up the rate of vertical travel during the stroke may cause the cylinder to become barrel shaped or bellmouthed because the hones are working longer in one area than another. A little experience and a steady, rhythmic

motion with the drill-press handle does the trick.

Start at the bottom of the cylinder where there is practically no wear. This is the area you will use as a centering means to reestablish a true cylinder bore.

1. Adjust the hones for a moderately firm contact down at the bottom of the cylinder so that they contact here but not higher up. The upper portion of the bore is worn to a greater diameter than the hones.
2. Start honing at the bottom and make tension adjustments as needed to maintain cutting. As you do this, the area of contact will gradually extend higher in the bore. Increase the travel of the hones so that you gradually contact the entire bore and the hone is emerging ½ inch or so at each end of the stroke.
3. You have now established a new, true bore, and the next step is to continue honing until you reach the desired oversize. Be sure to provide some means of dripping or squirting lubricant onto the hones while rotating them.

Finish-honing. When the bore has been trued over its entire length, it's time to measure with a micrometer to determine how close to, or far from,

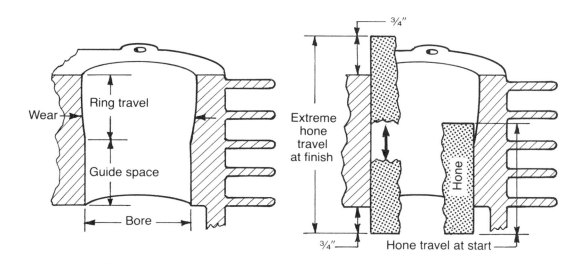

Since the lower end of the bore has little or no wear, use it to establish a center and a new bore dimension.

Place cylinder in a tub of hot soapy water and scrub vigorously, as shown above. You'll know you're finished scrubbing when you can wipe a clean paper towel firmly on the cylinder wall without showing any discoloration. Be sure all water is removed from pockets, passages, and bearings.

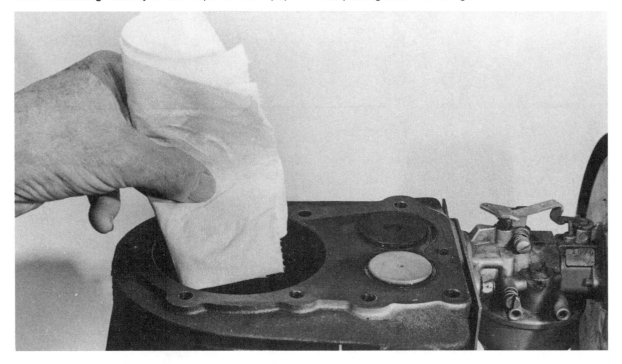

the desired finished oversize you are. Use caution and check often to avoid enlarging the bore too much.

When you've approached the final oversize bore diameter within about .0015 inch, replace the coarse hones with finishing hones and slowly work to the final dimension. Some mechanics advise going .0005 inch oversize so the finished dimension will be correct after the cylinder cools and contracts. This is acceptable practice if time is important and you have lots of experience. For a beginner, it's better to let the cylinder cool to room temperature as you approach the final measurement. If contraction results in a slight undersize, hone again, trying to avoid heating the cylinder. If this takes several cut-and-try steps, so be it. That's better than going too far.

Cleaning the bore. After finish-honing it is extremely important to thoroughly clean the cylinder bore. Even though it may appear clean and bright, be assured that there is plenty of honing swarf left in the minute hills and valleys of the crosshatch pattern. Flush, wipe, and bristle-brush the cylinder walls repeatedly with kerosene or solvent. This is a preliminary cleaning. Now place the cylinder in a bucket or laundry tub of hot water and soap. Use a round bristle brush to wash it several times, then dry it, making sure all water is removed from passages, pockets, and bearings. After drying, you should be able to wipe the bore with a clean, white handkerchief without picking up a trace of discoloration. Oil the freshly cleaned bore lightly to prevent rusting.

Many small engines will have a ball bearing pressed into place at the end of the crankcase opposite the removable plate. Even if you do not immerse the cylinder block completely, you'll probably get water into this bearing. Often that will be of no concern since you'll be replacing the bearing and seal anyway. If the bearing is still in good shape, however, blow out any water with compressed air and oil it liberally to prevent rusting and pitting. Also, look carefully for any oil passages and blow them clean of accumulated water and sludge. Use a small bristle brush to clean them as thoroughly as possible since you do not want honing debris to enter the bearings when it comes time for you to start your freshly overhauled engine.

19

Choosing and Installing a New Engine

The vast majority of small engine owners probably buy a piece of lawn or garden equipment with a suitable engine as part of the package. Many others, however, particularly those given to mechanical ingenuity, may elect to install an engine on almost anything from a log splitter, go-cart, or irrigation pump to a generator set. Still others may acquire a bargain piece of equipment with an engine not worth rebuilding, or may simply want to replace an existing engine with a more powerful or durable model.

All of that can be fun, but success will hinge on a considerable number of decisions. Among these are—

- The power and rpm required.
- Duty cycle and expected duration of service.
- Clutch and transmission used.
- How the engine will be mounted.
- Pulley sizes and retention.
- How to provide for engine cooling and service access.
- Whether to choose a two- or four-stroke engine, and whether it will have a vertical or horizontal crankshaft.

In short, you'll be doing some of the work normally done by the equipment manufacturer's engineers. Probably the worst-case situation is happening to have an engine on hand that was not originally intended for your application, but which you wish to cobble into a working power source. If you're lucky you may get by; if not, you'll find problem after problem.

HOW MUCH POWER?

Underestimating the power need is a common error. Just because an engine is labeled 3 hp or 5 hp does not mean that much power is available continuously to drive your machine. Although the adver-

tised power was indeed demonstrated on a dynamometer according to SAE procedures, it was done with a new engine built to laboratory standards, tuned for test, and corrected to standard sea-level barometric pressure and 60°F air temperature. If, however, you read the fine print on the manufacturer's specification sheets, you'll learn that production engines are still within limits at as little as 85 percent of their advertised bhp. Thus, there is a margin for manufacturing tolerances.

Moreover, after an engine is operated for a while, the buildup of deposits, valve and seat wear, and other factors will also diminish power slightly. Note, too, that *all* engine power tests are corrected to standard atmospheric conditions. Engine power can normally be expected to be reduced by 3 ½ percent for each 1,000 feet you are above sea level, and by 1 percent for each 10° above 60°F.

Consider also the fact that there can be major differences between power *developed* and power *delivered*. Almost any system of coupling to a load will introduce friction losses. That applies particu-

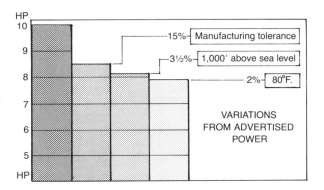

A new engine may produce as little as 85 percent of advertised horsepower.

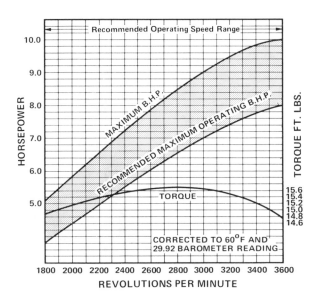

Standard SAE power curve shows horsepower and torque relative to engine rpm. Note the difference between maximum power and *recommended* maximum operating power.

power curves show two ratings: The upper one is the laboratory-test maximum horsepower, while the lower one is the maximum *recommended* operating horsepower.

Suppose, for example, that you plan to belt-drive an air compressor or orchard sprayer requiring 5 hp at full output delivery conditions. To that requirement you'd add 20 percent, or about 1 hp, for belt slippage and friction. Next, regard this 6 hp as about 60 percent of the expected engine horsepower. Thus, you're looking for a power curve that shows about 7.5 to 8 hp maximum for continuous use. Thus an engine in the 10-hp maximum rating bracket and an 8-hp maximum recommended operating power. For extended and heavy-duty use, think in terms of 80 percent of 8 hp—the recommended maximum operating power—which is a comfortable 6.4 hp. That may be unnecessarily conservative for a light-duty cycle and limited hours of usage, but it does allow for occasional overloads and normal wear over a period of time. It also allows the engine to run at less than maximum rpm, in this case at about 2,800 rpm.

larly to belts and gear drives. It is a rare transmission that is more than 80 percent efficient.

Overpowering. This can be an easy mistake to make, particularly if you have a sluggish piece of equipment or a larger engine sitting around. Though it may not hurt the engine, you'll probably find that the equipment belts, clutches, transmission, and bearings weren't designed to take the extra power.

Engineering Your Power

The best way to determine power needs is to consult the manufacturer's specifications for the equipment you plan to drive. If you have a hydraulic pump for a log splitter, an irrigation pump, an air compressor, a welder, or like machine, there's no question that the maker knows the power requirements.

You'll also need the specification sheets for likely engines in the power range involved, available from all engine builders. The sheets give power and torque curves relative to engine rpm, along with a wealth of other details. You'll find that the manufacturer's

Torque Requirements

As discussed in Chapter One, horsepower and torque are two very different things. Although it is customary to describe an engine as being rated at a given horsepower, it is important to note that the power curves show both horsepower and torque.

An engine can display remarkably high horsepower at very high rpm, yet fizzle out and die when heavily loaded. In an automobile, such an engine requires frequent attention to the gears to keep engine speed in an efficient range and have enough torque at the axles to provide performance. Many small two-stroke engines have such torque characteristics. When used on chainsaws, they are coupled to the load by centrifugal clutches that simply disengage if the speed drops too low.

Be sure to take a hard look at the torque curve of any engine you're considering and match it to the type of load you expect to apply. If you're building a log splitter, for example, you know that as the splitter blade encounters a tough log, the hydraulic-pump load increases rapidly and the engine tends to slow down. For this job you need an engine with a torque curve that builds up as the engine slows down. Such an engine might typically show a horsepower peak at 3,600 rpm, but the torque curve will

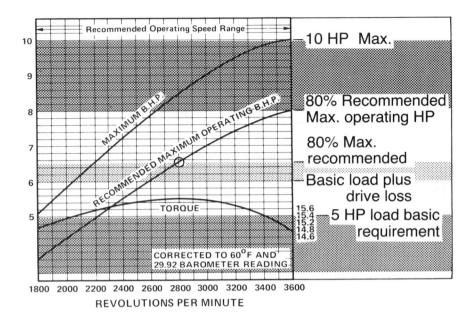

This power curve is used to select an engine to drive a load requiring 5 hp.

peak at about two-thirds that speed in the 2,400-to-2,600-rpm range.

Though you may not know, in foot-pounds, what the exact torque requirements of your job will be, remember that for jobs where the engine may be dragged down severely, the lower the peaking rpm for torque the better—unless you can provide a means of changing drive ratios as the engine slows down.

Durability

A 100 percent or even an 80 percent duty cycle means that the engine is working hard all or almost all the time. It calls for an engine of high durability, probably full cast-iron construction with heavy ball or roller main bearings and certainly with hardened valve faces and seat inserts.

At the other end of the scale, a light duty cycle means that although the engine works hard for a brief period, the intervals of light load allow the valves and seats, piston, and oil to cool. For such service it may be more economical to consider a more lightly constructed engine, although the engine must have the power to handle the load.

Converting from an electric motor. You may want to convert a machine originally driven by an electric motor to gas-engine power to gain mobility. Do not, however, be mislead by the electric-motor rating and expect to replace it with an equally rated gasoline engine.

Electric motors, especially those designed for industrial use such as for agricultural and construction machinery, are capable of handling heavy overloads for brief periods. A typical 1-hp electric motor cannot be replaced with a 1- or 2-hp gas engine. Check with the equipment builder, but expect to double or triple the power rating of the motor when converting to engine power. One rule of thumb you can count on is to provide at least 2 horsepower per kilowatt when powering a generator.

CONNECTING THE ENGINE TO THE LOAD

When you've managed to match up an engine to whatever it is you want to drive, the most critical factor is how you're going to connect the engine to the load. Only a very few types of loads can be

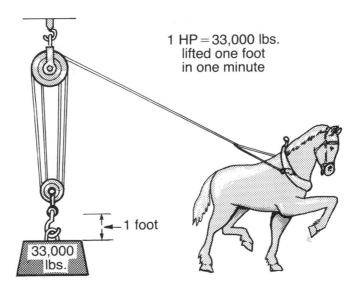

1 HP = 33,000 lbs. lifted one foot in one minute

1 foot

33,000 lbs.

Horsepower expresses the ability to do a certain amount of work in a given time, as shown above. The basic concept of torque, below, also carries over from the days of animal power. A horse was expected to dig in and increase his effort under load. We often expect an engine to do much the same thing.

Torque

connected directly without some means of clutching. Engines don't start if they are under load at the time of cranking. They must come up to speed and usually warm up a bit before they will pull a load without stalling. The exceptions are loads such as rotary mowers, water pumps, propellers or blowers, and generators in no-load mode.

Direct Connections

Even in the preceding examples of unclutched loads, direct, end-to-end connection of the engine crankshaft and the driven machine is a tricky business. Some generator and tractor manufacturers do insert the engine crankshaft directly into an armature or transmission shaft and bolt the engine face to face with these parts; but that arrangement is practical only under factory conditions where alignment and runout can be precisely controlled in the machining process.

Another serious concern with direct connection is harmonic or torsional vibration. Most driven equipment will have certain inherent vibrational characteristics from internal impellers, gears, electrical

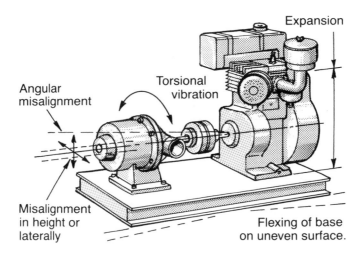

Direct, end-to-end coupling of an engine to a load will almost certainly cause trouble.

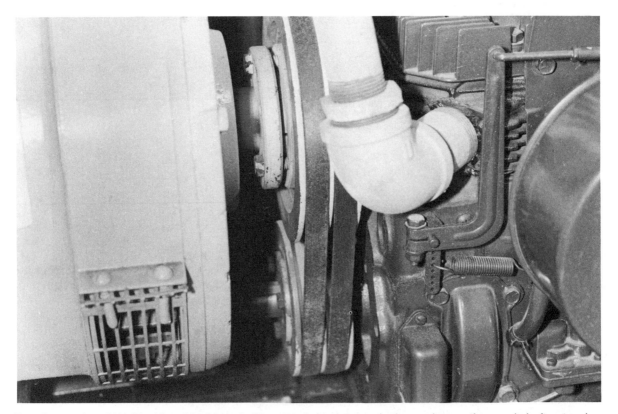

Two heavy-duty V-belts drive this home-built emergency generator. Note that the engine pulley is well back towards the engine on the crankshaft extension. This helps reduce harmful flexing forces.

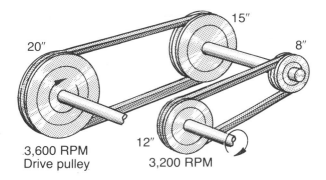

Jackshafts are often used to take off power at different speeds to multiple drives. The formula for figuring input-to-output speed on one like this is: 3,600 × 20/15 × 8/12 = 3,200 rpm.

poles, or compressor pulses. If these merge harmonically with the firing impulses of the engine, together they can cause vibrations strong enough to break the coupling or engine crankshaft or cause the flywheel key to shear. At best, you can wind up with a shaking, shuddering installation that can break fuel lines, gas-tank mounts, and blower housings.

If you feel such a layout is still for you, plan to use a semiflexible coupling with rubber or other resilient material between the two shafts. Although these, too, must be carefully aligned, they will soak up much of the vibration and help compensate for slight misalignment between engine and driven unit.

Belt Drives

Belt-driving remains one of the most attractive ways to connect a small engine to a load. Modern V-belts made for industrial drives are amazingly strong and durable, yet they offer enough elasticity to dampen engine vibrations.

Still another advantage of a belt drive is that it lets you select from a wide range of ratios to slow down or speed up the driven equipment. If your initial ratio doesn't work out quite right, it's relatively easy to alter the pulleys to suit your purpose. For example, you can easily gear an engine down to very slow speed but high torque for, say, a hoist in two or even three stages with a series of belts and pulleys commonly referred to as *jackshafts*.

Remember, however, to select a belt in the proper size and configuration to match the job. Don't try to use a similar-looking belt made for fractional-horse-power electric motors on a 5-hp engine. In addition to the increased load your gas engine imposes, a single-cylinder engine delivers a series of pulses reflecting each power stroke as compared to the smooth power of an electric motor. Those, too, will take their toll on a belt not designed to absorb them. You'll find that belt manufacturers list their products according to their proper applications and give recommended pulley diameters and V-groove widths. Belt widths are nominally 3/8 inch, 1/2 inch, and 21/32 inch in the range commonly used with small engines.

Belt adjustments. Unless you're planning to use a belt tensioner with a spring-loaded idler pulley, you must provide room for reducing the distance between pulleys to install the belt and then expanding it again to tension the belt. *Never try to install a belt by prying or stretching it over the pulley flanges.*

Tensioning adjustments may be slots for sliding the engine in the mounting bolt holes, jackscrews to raise the engine, or even a spreader screw or jack. Most belts are tightened until moderate thumb pressure between the pulleys will produce about 1/2 inch of deflection. If the belt must run tighter than this to prevent slippage, you should probably be using double-groove pulleys and a pair of matched belts.

A common pitfalls in tensioning belts is to become so involved with tensioning that you overlook the alignment of the two pulleys. The pulley faces must be parallel and in perfect alignment. Check this by placing a steel rule across the faces after tightening. It is very easy to skew the engine or driven shaft when shifting these parts in their mountings.

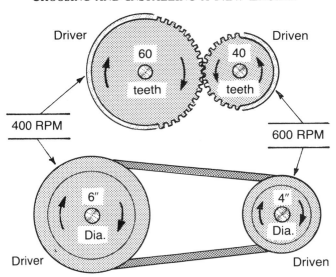

Gear rpm varies inversely with the number of teeth. Here, the tooth ratio is 6 to 4 and rpm ratio is 4 to 6.

Pulleys vary inversely with their diameters. Ratios are again 6 to 4 and 4 to 6.

Pulleys

A chronic problem with less expensive outdoor power equipment arises from the use of inadequate pulleys for V-belt drives. Usually these pulleys are made of pressed sheet steel or light zinc castings. Although they function fairly well when new, they often lose precision in their V-grooves so that the belts slip.

There is no point to building in such troubles if you're rigging your own engine drive. The first pulley you select must match the shaft configuration and keyway of your engine. The fit should always be snug and precise. Many larger hardware stores and industrial-supply houses display cast-iron pulleys with a variety of interchangeable hubs. You'll find such hubs are basically sleeves with a partial split. The outside diameter is slightly tapered.

To install these hubs you slip the inner hub over the shaft to the approximate correct position to align the belts. The outer portion of the pulley is then installed and, watching alignment as you go, the capscrews between the inner hub and pulley are tightened. This draws the tapered hub into the pulley and squeezes the hub down on the shaft and key very tightly. Such pulleys are machined true and balanced and will last practically forever. You should be able to find them in your hardware or farm supply store.

Pulley ratios. The pulley sizes you choose will determine the speed increase or reduction you get. Driving a smaller pulley with a larger one will speed up the driven equipment, whereas driving a larger pulley with a smaller one will slow it down. The ratio you select must be suited to engine speed and desired final drive speed. That may be a matter of convenience and space since the same ratios could be obtained with pulleys of several different sizes. Avoid extremes. It is not a good practice to use a very small pulley, especially on the engine, if you can work out the same ratio with a somewhat larger one. This has to do with something called *wrap area,* which simply means that a larger pulley provides more belt contact surface and thus doesn't need to be as tight. A larger pulley will also be less likely to slip and will put less flexing strain on the belt.

Another consideration in selecting pulley ratios is avoiding vibrations and harmonics. As an example, I designed and built the 5-kilowatt home emergency generator you see in some of the accompanying photos. The engine was rated at 16 hp and the generator speed for 60 Hz (cycles per second) was 1,800 rpm. I didn't want to run the engine at its full 3,600 rpm rated speed. Lower speeds were better for both increased engine life and lower noise level. In addition, 1,800 and 3,600 are even multiples and would almost certainly cause harmonic fighting between the engine and generator.

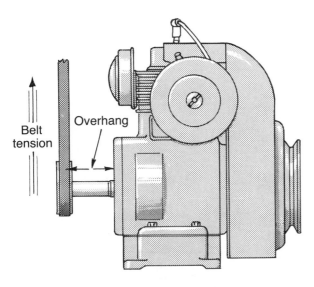

Excessive overhang, as shown here, can harm crankshaft bearings and belt. It can also set up flexing forces that shorten the life of the engine and driven parts.

The solution was to choose pulleys that allowed the engine to turn the generator at 1,800 rpm with an engine speed of about 2,750 rpm. That was also about the 10-hp point on the engine power curve—remember, 2 hp per kw. This worked out well and the unit runs smoothly without any visible flapping of belts, which would indicate a vibrational condition. Very few engine applications require that much

concern, but if you get into a problem, a change in ratios will usually get you out of it.

When installing a pulley on an engine crankshaft, or on any other shaft for that matter, do everything possible to minimize the overhang. Overhang is the distance between the pulley centerline and the front bearing of the engine. Keep the pulley as close to the engine as possible to reduce flexing load on the shaft and side load on the bearing.

Remember, the ratio of the pulleys is the same as the ratio of their diameters. If a 6-inch pulley is driving a 12-inch pulley, the ratio is 1:2. And if the 6-inch pulley is turning 2,000 rpm, a 12-inch pulley will turn 1,000 rpm.

Clutches

Since you will need some sort of clutch to disengage the engine for starting and other reasons on many applications, it is important to decide what is available and how you might apply it. Nearly all clutches, except dog-types, rely on bringing two friction surfaces together. The quality and type of clutch you select should be predicated on a number of considerations.

First, does the application call for frequent engaging and disengaging as is routine on a lawn tractor? Or, is it something you engage and leave engaged as long as the engine runs? Second, how much delicacy of control do you need? If a simple in-or-out, engaged-or-disengaged operation is adequate, that is one thing. But if you're driving a hoisting winch

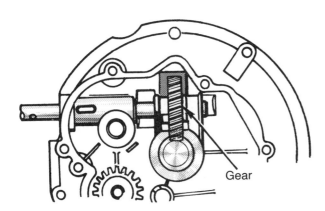

This gear reduction drives off a special crankshaft with a worm gear machined into the front extension. The

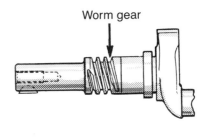

result is two output shafts with different speeds—one, perhaps, for a mower blade, the other for propulsion.

This centrifugal clutch drives the chain sprocket in a saw. When engine speed reaches proper rpm, the clutch members overcome the spring and move into contact with the drum's inner surface.

with your engine, you might want a clutch that permits you to slip it ever so gently to carefully manage a load.

Engines are available with a built-in gear reduction and clutch. The output shaft speed is reduced 8½ to 1 to the crankshaft speed. You can also buy wet- or dry-type clutches that mount directly on the engine crankcase but have no gear reductions. If your application requires industrial-level durability and the investment is justified, these units should be given serious consideration.

Centrifugal clutches. One of the smoothest and most pleasant clutches to work with is the centrifu-

gal type. These, of course, are widely used in chainsaws and small lawn machines.

In operation, at slow engine idle, the inner friction members, or *shoes,* retract from contact with the outer driven drum. Open the throttle, and as engine speed increases, the friction shoes move out under centrifugal force and seat against the drum to drive the chainsaw or whatever is involved.

Such clutches are available in many sizes and power ratings and are an excellent choice in the lower power ranges. As you get into higher horsepowers, however, the centrifugal action must be quite powerful to avoid burning the clutch surfaces.

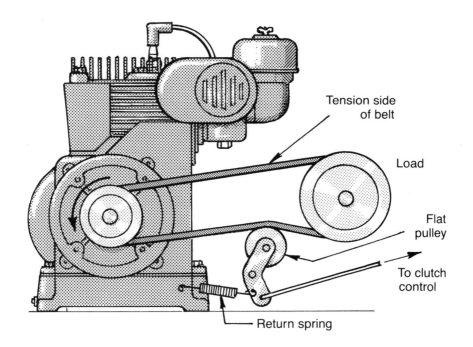

Try to put belt tighteners on the slack side of the belt so the pulling effort of the engine is not working against it. Also, keep the tightener on the outside of the belt to enhance friction contact.

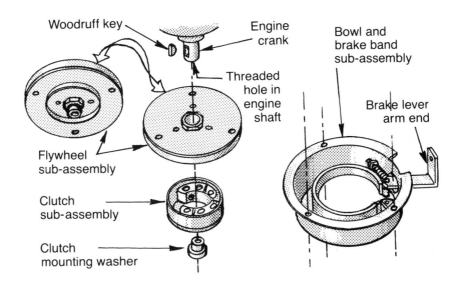

Blade-brake/clutches typically have a small auxilliary flywheel to compensate for the loss of flywheel effect when the rotary blade is stopped. Remember, the engine keeps running and must be balanced as before.

That means the springs that retract the friction shoes must be set up so they do not allow engagement until the engine reaches a rather high speed where the centrifugal action is strong. This tends to limit any maneuvering of the load with the engine just above idle speed, and as a result the engagement may be abrupt.

Belt-tightener clutches. Nearly all rider and self-propelled mowers, lawn tractors, and similar lawn and garden machines use some form of belt-tightener clutches. Such clutches rely on the friction between the V-belt and drive pulley. Loosen the belt enough and it slips; tighten it and it drives. This is not the best arrangement but it is inexpensive, easy to maintain, and it works.

A belt-tightener clutch ordinarily consists of a flat pulley mounted on a swinging lever, which provides the operator with the leverage to force the pulley against the back, flat surface, of the belt. The pulley, or *idler,* should have ball or roller bearings to minimize friction and increase durability. Although such pulleys are often used with so-called permanently lubricated bearings, it is better to use the type with a grease fitting if you anticipate extensive equipment use or long periods of storage.

If the machine requires slipping the clutch, as when easing a tractor into a parking spot, the tightener can be arranged so the operator must maintain hand or foot pressure on it at all times to maintain engagement. If the tightener is simply for in-and-out use, a locking device can be used to hold the belt tightener. If you rig your own belt tightener, the idler pulley should be on the side of the belt that returns to the power source. Thus, the tension side of the belt is not deflected.

Blade-brake/clutches. In Chapter Nine I discussed rotary-mower engine brakes. These devices are definitely part of the engine and are built by the engine manufacturers. All of them stop *both* the engine and the blade when the operator's control is released. A blade-brake/clutch is quite different and is much more convenient for the operator since only the blade, not the engine, stops.

Blade-brake/clutches are not part of the engine but have a definite effect on engine performance. Some of the original models were not satisfactory because of engine problems such as stalling. If you have one of those, you may have been puzzled by what appeared to be engine problems.

There were two basic problems with the early

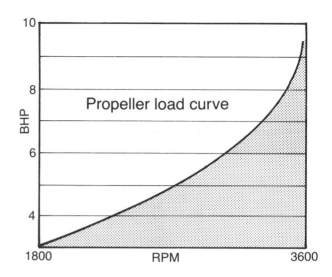

Pumps that handle air or liquid absorb power along a propeller-load or *cube* curve in which the load increases as the cube of rotational speed.

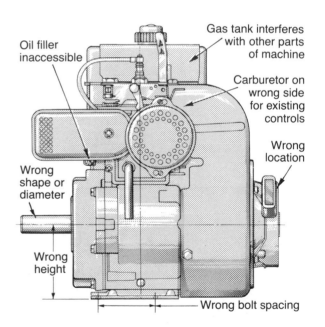

Shown above are typical detail differences you may encounter when engine swapping. Tecumseh's installation drawing on facing page includes the location of major parts and connections, mounting-bolt placements, and other vital details.

blade-brake/clutches, which simply de-clutched the blade from the engine and braked the blade to leave the engine running. First, the engine builder, not necessarily anticipating a blade-brake/clutch, planned on the inertia of the rotating blade serving as a form of flywheel mass. When the clutch disconnected the blade from the engine, the built-in flywheel mass was a bit on the light side. When the clutch was re-engaged and tried to start the blade spinning, the engine stalled—especially if the blade happened to be bedded in heavy grass.

The second factor was that the engagement of the clutch dropped an instantaneous load on the engine, which was running at high idle throttle with the throttle valve nearly closed. That meant the governor had to act very quickly to open the throttle and prevent stalling. In many cases the simple air-vane governor wasn't fast enough, and the carburetor didn't supply the extra fuel in time. Together, these two factors resulted in frequent engine stalls.

More recently, blade-brake/clutches have been built with a small flywheel as part of the assembly to prevent those problems. In fact, in some cases the extra mass helps the mower handle heavy cutting loads better.

From the standpoint of owner service, about all

Tecumseh Installation Dimensions

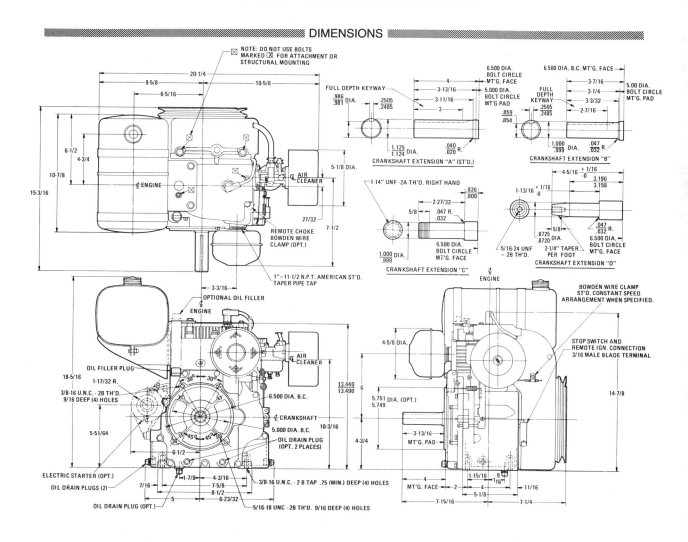

that you can do is make certain the control linkage to the operating arm on the mower deck hasn't stretched or become kinked, and that the control-arm travel is adequate. The operation of the clutch and brake can be checked manually. Disconnect the spark plug and try turning the blade in both the running and braked mode. I do not recommend attempting to disassemble a blade-brake/clutch. If you have trouble, this is a job for a dealer.

Gear Drives

Unless you plan to salvage a transmission from a used tractor or the like, gear drives are seldom a practical solution to a home-engineered, small-engine application. One exception might be a worm-gear speed reducer. Industrial-supply houses carry them in various ratios.

NON-MECHANICAL LOADS

So far I've been talking primarily about the type of machinery where you might have a belt, gear transmission, straight coupling, or the like driving a mechanical load with wheels, tines, choppers, or other solid parts doing the work. Not all loads fit this picture.

Examples of non-mechanical loads are a centrifugal water pump supplying a sprinkler system or moving irrigation water. Such a pump, together with marine propellers and fixed-pitch aircraft propellers, absorbs power pretty much along a curve sometimes called a *propeller-load curve.* The power absorbed increases as the cube of the rotational speed. This means that the load increases with speed, slowly at first and then climbing steeply as the speed goes up until finally the curve becomes almost vertical.

Such loads tend to be self-limiting. Most marine propellers are the result of "cut and try." If the initial pitch or diameter are off only a trifle, the engine will simply never reach the design speed but will level off at full throttle and be seriously overloaded for continuous service. The same principle applies to pumps, blowers, and similar non-mechanical loads. This peculiar form of load provides a kind of built-in power control. If the back pressure on a pump is increased, the pump will slow down until an rpm is reached where the engine can handle the load at a constant speed.

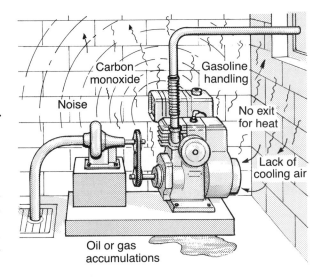

Carbon monoxide · Noise · Gasoline handling · No exit for heat · Lack of cooling air · Oil or gas accumulations

Although locating equipment such as this engine-driven sump pump in a basement often seems attractive, such installations are fraught with hazards as shown.

AC Generator Loads

Alternating-current generators are another type of special load. Since the rotating poles of the armature must cut the magnetic lines of force at exactly the right speed to produce 60-Hz power, the engine cannot be allowed to slow down under load to build up torque. The only time this will happen is under overload, at which point the frequency and voltage will drop abruptly and a fuse or circuit breaker should blow.

Thus, if you're choosing an engine for a generator set, you must have sufficient power instantly available to pick up instantaneous loads. An example is when your generator is running no-load and you throw the transfer switch to pick up your household load. In short, the engine must operate under all normal loads at the correct governed speed but have plenty of available throttle left to handle a motor cutting onto the line. An engine that is barely capable of handling constant electrical loads will not be able to cope with instantaneous loads, particularly motor-starting loads. That's one reason for the rule, of at least 2 horsepower per kilowatt.

Either of these home-built generators can be moved about for service, plugged into the transfer switch on the wall, and connected to the muffler outside for clean, hazard-free operation.

ENGINE MOUNTING AND INSTALLATION

Although it might be nice in some cases to mount a small engine on rubber mounts or even stiff springs to isolate vibration, this does not work out quite as well with single-cylinder engines as it does with automotive engines.

The nature of the single-cylinder engine is not inherently smooth, even with balancing, so such mounts tend to introduce substantial movement. Since the connection to the driven machinery is normally critical, such movement is a problem and the only solution might be to introduce the complication of a universal joint like those used in automotive drivelines. U-joints are often used in mower-deck and snowblower drives, but for the reason that the implement must be moveable relative to the engine and transmission.

Therefore, most small engines are rigidly mounted with bolts into the engine base or crankcase face. A substantial mount helps to minimize vibration and resist belt tensions. Take precautions such as effective lockwashers or safety wiring to prevent the bolts from backing off and dropping out. If you're using a belt drive, this may be the place to provide some slots to allow moving the engine for belt take-up.

Engine Placement

There are other things to consider before you weld in your engine mounts. For example, will your chosen location be directly exposed to a constant flow of grass cuttings, chaff, or sawdust? Can you easily remove the blower housing for cleaning the cooling fins? Is the blower housing too close to a wall or some other barrier that might block cooling air flow? Many builders of home equipment like to enclose the engine and sometimes forget that it's air-cooled.

For the same reason, never locate engine-driven equipment in an enclosed space. Many times it appears attractive to locate an emergency generator or an emergency sump pump in a small basement room or partitioned-off area of a garage. There are three good reasons not to do that.

First, there is always a danger of carbon monoxide infiltrating the rest of the house. Second, there is the possibility of fire with a gasoline-powered engine. And third, the engine will overheat.

Remember, an engine rejects about 75 percent of the heat energy provided to it as radiant and exhaust heat. For example, if you were to locate a 10-hp-engine generator in an 8-by-10-foot enclosed room the room would become so hot within about 20 minutes that the engine couldn't cool and would probably seize up.

Providing cooling air. One way you can locate an engine in its own room is to provide a set of louver doors or other quickly and easily opened air access. It may also be necessary to install a means of exit for the heated air. And, you'll need an exhaust system with a flexible connection to discharge the exhaust outside.

I have two home-built 5kv emergency generators. One is powered by a four-cylinder, water-cooled engine, the other by a single-cylinder, air-cooled engine. I could have provided some sort of shed or housing outside, but that would probably have meant digging through snow or fighting a blizzard on a night when I needed power the most. I could have installed them in my well-pump room or my darkroom, both separate rooms off my basement. While that would have kept the engines warm and dry for easy starting, both units are extremely heavy and trying to get them to the basement, or out if service was needed, posed a formidable problem.

The only other choice was a heated and attached garage. Here, I could locate the generators near the power distribution panel and wire in a transfer switch directly off the panel. I could also place the starting battery near a charger. The garage walls are concrete block and the ceiling is plaster over metal lath. The only access door to the house is an approved metal fire door with a fusible link and gravity closure, so the fire hazard is low.

There are other benefits. If the starting battery fails, I have two cars and a tractor, each with a battery that can be jumped to start the engines. Opening a garage door allows plenty of ventilation, and an exhaust connection runs through the wall to an oversize muffler outside.

Finally, both units are mounted on heavy-duty industrial castors that allow me to move them around so I can get at either side for service. If there's a problem with one, I can wheel it aside and roll the other in and connect the exhaust pipe with a coupling provided for the purpose.

Remote fueling. As a further effort to minimize the fire hazard, the single-cylinder engine that is the prime unit never has over a pint of gasoline in the tank for starting purposes. A 6-gallon tank, fitted with a hose and outboard-motor-type connection fittings, is the main fuel supply and it's kept away from the engine. If refueling is necessary, I can switch back to the engine tank for a minute or two while refilling the remote tank outside.

AUTOMATIC STARTING

Every year a great number of people are left without power because of natural disasters or for other reasons. Apparently many of them dream of an emergency, engine-driven generator that will start and switch on automatically. Many who have asked my advice on this are disappointed when my answer is, "Don't even try it."

Starting an engine automatically is a complex and costly procedure, and—short of almost daily inspection—not very reliable. For one thing it requires at least one starting battery in top condition, which in turn requires some form of almost constant trickle charging. To start an engine automatically you need a number of electrical controls and devices in addition to the starter. Briefly, here's the sequence of events necessary to start your engine and get it on the line:

- Initial power failure. Is it momentary or extended? You don't want your engine trying to start every time the lights flicker or go off for a few seconds. A timing sensor must be energized and monitor the line for a minute or two before initiating a start.
- Open the feed circuit from the utility lines. When your engine starts you don't want power feeding back out to be a hazard to a utility crew or perhaps to feed your neighbor's house.
- Open a battery-operated fuel supply valve. If the tank is remote, a fuel pump must be started to supply the carburetor.
- Close an electric choke and set the throttle. This must either be timed out or sequenced to the fuel pump system.
- Initiate cranking and sense the cranking period. If the engine doesn't start, the cranking system must shut down and allow the starter to cool.
- Repeat starting attempt. If the engine starts, the choke must be backed off in sequenced, timed steps.
- Sense engine warmup and temperature.
- Sense engine rpm and generator output voltage and frequency.

- Close the transfer switch to your household circuits.
- Open motorized cooling air vents or doors.

After all of this happens, successfully, you should now have your lights on and everything running. But that's the problem. Your little engine will also be trying to feed a multitude of loads that may be on or off randomly. For example your freezer, refrigerator, well pump, furnace blower, water heater, maybe a heat pump or air conditioner—plus whatever else you have—can all be grasping for power simultaneously. In many modern homes that calls for about 20kw, or an engine of about 35 to 40 hp. Suddenly, we're not talking about a small gas engine. The only option is to arrange your transfer switch to feed only selected emergency circuits, but that requires rewiring your house and distribution panel.

It's only fair to ask how I manage on a 5kw standby unit. This is something you can do and it's not difficult. When the power fails, I take a flashlight and pull the fuses or switches on all but a few lighting circuits. I always leave the circuit to the garage lights on.

I then open the fuel valve, connect the battery, choke the engine manually, and push the starter button. After the engine warms up and stabilizes I manually throw the transfer switch and the garage lights come on. This is the time to switch to the main fuel tank. Then, depending on what's important at the moment, I restore power to one or two circuits, often for brief periods. Our home has operated for up to four days at a time that way.

Moral: *Small engines are wonderful things and will do the job if you treat them right and give them a chance.*

Index

Air filter:
  cleaning, changing of, 26
  removal from storage, 37–38
Air vanes and weights (governors), 98–100
Alternators:
  functions of, 148
  servicing of, 160–167
Ammeter, 155
Antique engines, 277–278
Armature:
  air-gap adjustment, 136
  service of, 159
Automatic compression release, 250–252
Automatic spark advance, 252

Baffle and loop scavenging, 11
Ball and roller bearing, 234–241
Ball-type clutch starter, 58–60
Battery:
  function of, 149–150
  generator and, 146
  storage of, 34–35, 36
Battery connections:
  alternator service, 161
  starting problems, 44
Bearings. See Engine bearings; Generator bearing service; *entries under names of specific bearings*
Belt drives, 295–296
Brake. See Flywheel brake
Breaker-point system and service, 125–141
  cleaning and inspection of, 126–129
  coil checking, 135–136
  condenser testing, 133–134
  described, 122–125
  external breakers, 132–133
  ignition timing (external breakers), 137–140
  ignition timing (internal breakers), 136–137
  impulse magneto timing, 140–141
  location, 125–126

replacement, 129–132
timing two-stroke engines, 141
Brushes (generator):
  functions of, 147–148
  replacement of, 159–160
  service of, 158–159
Bushing-type bearing replacement, 234
Butane fuel systems, 95–96

Camshaft, 245–250
  automatic compression releases, 250–252
  automatic spark advance, 252
  inspection of, 246–248
  problems with, 245–246
  timing of, 248–250
Capacitor-discharge ignition, 143–145. *See also* Solid-state ignition systems
Carbon removal, 31–32
Carburetor, 77–93
  cleaning of, 55–57
  diaphragm carburetor, 87–91
  float-bowl carburetor, 77–87
  function of, 77
  inspection of, 53, 55
  tank-top carburetor, 91–93
  two-stroke engines, 14
  See also Fuel system and service
Carburetor chokes. See Chokes
Chainsaw switches, 52
Chokes, 71–77
  full thermostatic choke, 74–75
  function of, 71–72
  inspection of, 53
  mechanical-link chokes, 72–73
  primers, 75, 77
  removal from storage, 37–38
  vacuum chokes, 73–74
Clutches:
  engine load connection, 297–302
  See also Manual starters
Coil:
  ignition-spark requirement, 120–121

operation of, 125
testing of, 135–136
Cold-weather starting, 43–44
Combustion, 7–8
Commutator:
  function of, 147–148
  service of, 158–159
Compression and compression stroke:
  four-stroke engines, 5, 6
  starting problems, 47, 57
  two-stroke engines, 11
Condenser:
  breaker-point system, 123
  testing of, 133–134
Connecting-rod bolts, 259–261
Connecting rods:
  piston removal, 258–259
  torquing of, 174–177
Consumer Product Safety Commission, 110
Continuity checks, 153
Controls, 27–28
Cooling fins, 25–26
Coverings (protective), 33
Crankcase, 221–243
  bearing repair/replacement, 233–243
  bearing types, 231–232
  components and fits of, 224–231
  function of, 221–224
Crankcase sealing, 14
Crankshaft:
  balancing systems, 252–254
  inspection of, 244–245
Crankshaft end play, 225–226
Cylinder block cleaning, 191
Cylinder bore, 279–289
  breaking in of, 280–282
  function of, 279
  inspection of, 282–284
  materials for, 279
  oversizing, 284–289
Cylinder heads, 187–198
  compression releases, 198
  cooling of, 187

Cylinder heads (continued)
    head gasket installation, 191–192
    removal/inspection of, 187–198
    torquing down the head, 192–
        198
    types of, 187
Cylinder wall protection, 29

Diaphragm carburetors, 87–91
Diaphragm/pulse fuel pump, 71
Diode testing, 165–166
Dog-type clutch starter service, 61,
    63
Driveshaft seals, 227–231

Electrical switches, 35–36
Electrical system and service, 146–
    172
    alternators, 160–167
    components of, 146–152
    diagnostic tools in, 153–155
    generators, 156–160
    starter, 170–172
    voltage regulator (mechanical),
        167–170
Electric-drill starting, 67–68
Electric starters, 150–151
Emergency generator:
    engine selection for, 302
    fuel for, 69
    governors for, 107–109
    storage of, 30
Engine bearings:
    ball and roller, 234–241
    bushing-type, 234
    needle-type, 241–243
    repair/replacement, 233–243
    taper-bearing end play, 226–227
Engine cleaning (exterior), 29
Engine oil change, 28. See also Lu-
    brication system; Oil
Engine selection (new), 290–305
    connecting engine to load, 292–
        302
    mounting and installation, 303–
        304
    non-mechanical loads, 302
    power requirements, 290–292
    starting, 304–305
Exhaust stroke, 5, 6, 7

Factory specifications, 179–180
Float-bowl carburetor, 77–87
Flywheel, 110–119
    and breaker-point system service,
        125
    four-stroke engine, 5, 7
    functions of, 110–111
    ignition-spark and, 120–122
    inspection of, 112–115
    removal of, 111–112
    requirements of, 110
    ring-gear service, 118–119
    two-stroke engine, 12
Flywheel brake:
    inspection of, 115–118
    purpose of, 110–111
Flywheel-fan service, 119
Four-stroke engines, 3–8
    combustion essentials, 7–8
    compression stroke, 5, 6
    crankcase, 221
    exhaust stroke, 5, 6, 7
    intake stroke, 3, 5, 6
    operation of, 3–7
    piston installation, 272–273
    power stroke, 5, 6
    starting problems, 41–42
    storage of, 25–30
    two-stroke engine compared, 7, 8
Fuel:
    protective fuel mix, 29, 30
    winter/summer mixes, 69
Fuel-and-oil lubrication, 13
Fuel leaks, 37
Fuel line:
    carburetor types, 77–93
    checking of, 35
    inspection of, 54
Fuel pump:
    diaphragm/pulse type, 71
    inspection of, 54
    mechanical, 70–71
Fuel system and service, 69–96
    carburetor chokes, 71–77
    fuel pump (diaphragm/pulse
        pumps), 71
    fuel pump (mechanical), 70–71
    natural gas systems, 93–95
    propane/butane systems, 95–96
    protection of, 29

starting problems, 45, 47, 53–57
    two-stroke engine storage, 31
    winter/summer fuel, 69
    See also Carburetor
Fuel tank:
    checking of, 34
    draining of, 28–29
Full thermostatic choke, 74–75
Fuses (electrical), 152

Gaskets:
    crankcase, 224–231
    See also Head gasket
Gasoline. See entries under Fuel
Gear-drive clutch starter service, 63
Gear drives, 302
Generator:
    function of, 146–148
    servicing of, 156–160
Generator bearing service, 159
Generators (emergency). See Emer-
    gency generator
Generator/starter motor service,
    156–160
Governors, 97–109
    air vanes and weights, 98–100
    function of, 97–98
    linkages, 101–104
    operation of, 98–104
    servicing vane-type governors,
        107–109
    servicing weight-type governors,
        104–107
    springs, 100–101
Governor springs, 100–101

Head gasket:
    inspection of, 57
    installation of, 191–192
    See also Gaskets
High-voltage coil. See Coil
Honing:
    cylinder bore, 280–282
    cylinder bore oversizing, 287–289

Ignition switch, 47
Ignition system and service, 120–145
    breaker-point system, 122–125
    breaker-point system service, 125–
        141

ignition-spark requirement, 120–122

solid-state systems, 141–145

starting problems, 45, 47–53

See also Breaker-point system and service; Manual starters; Solid state ignition systems; Starter; Starting problems

Intake stroke:

four-stroke engine, 3, 5, 6

two-stroke engine, 9

Interlocks, 52–53

Linkages (governors), 101–104

Lubrication system:

described, 254–256

four-stroke engine, 7

See also Oil

Magneto:

flywheel and, 120–121

timing, 140–141

unit-type magneto, 124–125

Maintenance. *See* Seasonal and routine service

Manual rotation, 27

Manual starters, 58–68

ball-type clutches, 58–60

checking of, 26–27

dog-type clutches, 61, 63

electric/manual start kits, 67–68

gear-drive clutches, 63

general service techniques, 58

pawl-type clutches, 61, 62

recoil spring rewinding, 66–67

rewind-spring replacement, 67

starter rope replacement, 64–66

See also Ignition system and service; Starter; Starting problems

Mechanical fuel pump, 70–71

Mechanical-link chokes, 72–73

Microchip replacement kits, 145

Mufflers, 12

Natural gas fuel systems, 93–95

Needle bearing replacement, 241–243

New engine selection. *See* Engine selection (new)

Nicad batteries, 150

Ohms measurement, 154–155

Oil:

changing of, 28

cold-weather starting, 43–44

lubrication systems, 254–256

removal from storage, 37

See also Lubrication system

Overhead-valve engines, 215–220

Pawl-type starter service, 61, 62

Piston pin-bore inspection, 264–265

Piston-pin removal, 261–264

Piston porting, 10, 11

Piston rings:

cleaning of, 266–268

groove wear estimation, 268–269

installation of, 270–277

removal of, 266

Pistons, 257–269

inspection of, 265–269

installation of, 270–277

installation of (four-stroke engine), 272–273

installation of (two-stroke engine), 273–274

removal of, 258–265

torquing connecting rods, 274–277

variations among, 257–258

Plug fouling. *See* Spark plug; Spark-plug fouling

Points. *See* Breaker-point system and service

Polarization, 160

Power stroke, 5, 6

Primary wire problems, 51–52

Primers (chokes), 75, 77

Propane fuel systems, 95–96

Protective coverings, 33

Protective fuel mix:

emergency generators, 30

storage, 29

Pulleys, 296–297

Pumps, 302. *See also* Fuel pump

Rebuilding considerations, 175–186

diagnosis of problem, 178–179

rationale for, 175–178

service bulletins/factory specifications, 179–180

short block purchase, 175

tools and, 23, 175

wear estimation, 180–185

working parts versus running parts, 185–186

Recoil spring, 66–67

Recoil starters. *See* Manual starters

Reed valves:

automatic compression releases, 250–252

protection of, 33

two-stroke engines, 10, 11, 220

Regulator. *See* Voltage regulator

Relays, 151–152

Remote switches, 52–53

Repair. *See* Seasonal and routine service

Rewind-spring replacement, 67

Rod-cap matching, 259

Rope. *See* Starter rope

Routine service. *See* Seasonal and routine service

Rpms, 13–14

Seals:

driveshaft, 227–231

inspection of, 32–33

See also Gaskets; Head gasket

Seasonal and routine service, 24–38

four-stroke storage, 25–30

non-engine parts, 34–36

removal from storage, 36–38

storage damage, 24

two-stroke engines, 30–33

Service bulletins, 179–180

Short block, 175

Solenoids, 151–152

Solid-state alternator service, 163–164

Solid state ignition systems:

advantages of, 141–142

capacitor-discharge ignition, 143–145

components of, 142–143

described, 141

servicing of, 145

See also Ignition system and service

Spark:

armature air-gap adjustment, 136

ignition system servicing, 120–122

Spark (continued)
  See also Breaker point system and service
Spark advance, 5
Spark plug:
  cylinder head inspection, 190–191
  reading of, 46–47
  removal from storage, 38
  replacement of, 26
  service of, 47–51
Spark-plug fouling, 14
Starter:
  flywheel ring-gear service and, 118–119
  troubleshooting, 170–172
  See also Manual starters
Starter motor. See Generator/starter motor
Starter rope, 64–66
Starting fluid, 44
Starting problems, 41–57
  battery connections, 44
  cold-weather starting, 43–44
  compression checking, 47, 57
  flooded engine, 43
  four-stroke starting, 41–42
  fuel system troubleshooting, 45, 47, 53–57
  head gasket leakage, 187
  heat application for, 44
  hot engine, 42
  ignition troubleshooting, 45, 47–53
  motor/generator service, 156–157
  spark plug service, 47–51
  starting fluid use, 44
  See also Manual starters
Storage:
  damage from, 24
  four-stroke engines, 25–30
  non-engine parts, 34–36
  two-stroke engines, 30–33
Switches:
  function of, 151
  protection of, 35–36
  starting problems, 51–53

Tank-top carburetor:
  described, 91–93
  inspection of, 55
Taper-bearing end play, 226–227
Thermostatic choke (full), 74–75
Throttle, 37–38
Timing:
  automatic spark advance, 252
  camshaft, 248–250
  crankshaft balancing weights, 253–254
  external breakers, 137–140
  impulse magnetos, 140–141
  internal breakers, 136–137
  two-stroke engines, 141
Tools, 15–23
  diagnostic (electrical system), 153–155
  rebuilding, 23, 175
  routine service, 15–22
Top Dead Center (TDC), 3, 5, 7, 8
Torque:
  cylinder heads, 192–198
  engine selection, 291–292
  four-stroke engine, 8
  two-stroke engine, 14
Two-stroke engine, 9–14
  advantages of, 9, 12–14
  automatic compression releases, 250–252
  characteristics of, 12–14
  compression in, 11
  disadvantages of, 14
  four-stroke engine contrasted, 7, 8
  intake in, 9
  operation of, 9–11
  piston installation in, 273–274
  reed valve service, 220
  scavenging cycle, 10–11
  storage of, 30–33
  timing of, 141
  transfer action in, 9–10

Vacuum chokes, 73–74
Valve clearance, 211–215
Valve guides, 200–201

Valves, 199–220
  closing of, 29–30
  cylinder head inspection, 190
  function of, 199–201
  inspection of, 57
  reed-valve service, 220
  servicing of, 202–209
  valve train overhaul (L-Head), 201–215
  valve train overhaul (overhead-valve engine), 215–220
Valve seats:
  function of, 200
  servicing of, 202–209
Valve springs:
  function of, 201
  inspection of, 209–210
Valve tappets, 246–248
Valve train:
  four-stroke engine, 7–8
  overhaul (L-Head), 201–215
  overhaul (overhead-valve engines), 215–220
  reassembly, 210–211
Vane-type governors, 107–109
Voltage measurement, 153–154
Voltage regulator:
  function of, 149
  service of, 167–170
Volt/ohmmeter (VOM):
  alternator service, 160–161, 162, 163
  coil testing, 135–136
  condenser testing, 134
  diode testing, 165–166
  motor/generator service, 157
  starter troubleshooting, 170, 172
  use of, 153–155
  voltage regulator testing, 168

Wear estimation, 180–185
Weather:
  gasoline, 69
  starting problems, 43–44
Weight-type governors, 107–109
Wiring problems, 51–52

For information on how you can have **Better Homes and Gardens** delivered to your door, write to Mr. Robert Austin, P.O. Box 4536, Des Moines, IA 50336.